ROBERT W. DEUTSCH

NUCLEAR POWER

A RATIONAL APPROACH

FOURTH EDITION

GP COURSEWARE
Columbia, Maryland

GP Courseware, Inc., Columbia, MD 21044

Published 1975. Fourth Edition, 1987
Printed in the United States of America
93 92 91 90 89 88 87 5 4 3 2 1

Library of Congress Cataloging-in-Publication Data

Deutsch, Robert W., 1924-
Nuclear Power.

1. Nuclear power plants. I. Title.
TK9146.D48 1987 363.1'79 86-29444
ISBN 0-87683-831-X (pbk.)

Contents

Preface

Nuclear power plants have been misnamed. Since the heat source used in these plants is uranium that is directly fissionable, or uranium that has been converted to a fissionable form, a more appropriate name would be uranium power plants. This name compares directly with fossil power plants that burn coal, oil, or gas as the heat source.

Of course, a uranium power plant would still be identical in every respect to a nuclear power plant. However, the word "nuclear" is associated with weapons in most people's minds. Even though nuclear power and nuclear weapons are two distinct and different technologies, no matter how rational the discussion, sooner or later words such as proliferation and plutonium enter the discussion and the distinction between the technologies tends to disappear. As an example, China has been exploding nuclear bombs for years but is just starting to look for assistance from other countries including the United States to build uranium or nuclear power plants that generate electricity.

Shakespeare wrote, "What's in a name? That which we call a rose by any other name would smell as sweet." In the case of power plants, a nuclear power plant by any other name would be more acceptable to the public. Unfortunately, the word "nuclear" is here to stay. This book should give the reader a different perspective to help dispel the irrational fear that may be conjured up by the word "nuclear."

The statistics presented in this book are derived from reliable sources. The information relating to all facets of the nuclear power controversy is well known. Those in favor of nuclear power consider this information sufficient justification for building more plants, while those opposed point to the same information as reason for terminating the program.

About the Author

Dr. Robert W. Deutsch is Chairman of the Board and President of General Physics Corporation, a company he founded in 1966. Under his leadership, the Company has grown from an organization that provides training and support services to the nuclear utility industry into a total service company dedicated to more efficient operation and maintenance of all types of capital-intensive facilities.

Prior to General Physics, he served for eight years as Chairman and Professor of the Department of Nuclear Science and Engineering, The Catholic University of America, in Washington, D.C. He received a B.S. degree in physics from the Massachusetts Institute of Technology and a Ph.D. in physics from the University of California at Berkeley. He is a registered Professional Engineer in Maryland and Michigan.

1 Fear and Nuclear Power

The introduction of new technology has always created fear and apprehension among users and observers. Electricity and natural gas caused similar apprehension, as did the train, the automobile, and the airplane. The fear of the new and different tends to disappear as people become accustomed to the new technologies. Although all these technologies can and do cause death and destruction, the general population is willing to accept man-made disasters as long as the frequency is reasonable and the benefits outweigh the risks. More than 4,000 people are killed in the United States each year by electricity. Yet it is doubtful if anyone in today's society considers turning on a light switch a high-risk action.

The 1979 accident at the Three Mile Island Nuclear Plant (discussed in Chaper 6) caused no deaths or injuries and will have no effect on future generations, but the fear instilled in the public by the news media made it appear that a major catastrophe had occurred. Furthermore, to perpetuate the myth of tragedy, the news media sponsors an anniversary dedicated to the "survivors" of Three Mile Island to celebrate what could have happened rather than what did happen. On the other hand, it is well known that the smoking of cigarettes and the drinking of alcohol are a major cause of birth defects, and the long-term consequences of these activities result in more than 350,000 deaths each year. This is a real, not an imagined, catastrophe that threatens the health and curtails the life expectancy of the young and impressionable.

Fear gets people's attention. Prediction of an impending disaster, no matter how small the probability, is much more newsworthy than reassuring the public that the real dangers are small. Even though the predictions of catastrophe prove to be grossly in error, the fear generated lingers on. This is particularly true if the news media continues speculating on why their initial predictions went astray, attempting to convince the public that it was lucky this time but perhaps not the next.

Dr. Robert L. Dupont, a nationally known practicing psychiatrist with wide experience in both print and broadcast media, analyzed an aggregate of thirteen hours of videotaped news broadcasts on nuclear power. His scrutiny of the topic led him to observe that the major television networks have injected fear, intentionally or otherwise, into an already fear-inspiring subject.

Dr. Dupont notes that nuclear opponents cleverly restrict the discussion of nuclear power to the supposed dangers for the purpose of instilling fear and indicating that the only solution is totally eliminating nuclear power. Applying the same line of reasoning to other accepted activities, he states,

> If all you have to know about the automobile is that you can get killed by it, then you do away with the automobile. Or, if all you have to know about the airplane is that it can crash and kill you and that once you are on that plane there is nothing you can do to protect yourself, you would not go up in the air. If all you had to know about food is that you can get fat and have diabetes and heart attacks, you would, I suppose, do away with food. This is a very strange way to look at any issue. You can understand how absurd this position is if you think about water instead of nuclear energy. I can see the television picture closing in on a swimming pool, and the voice-over saying, "Do you realize that there is enough water in this pool to drown 100,000 people?" It is true, and it is the kind of statement we often hear today about nuclear energy. All these statements magnify our fears and remove meaningful perspective. This is a powerful technique of persuasion.

The technical and economic issues associated with the nuclear power program can be addressed to the satisfaction of the American public. However, the emotional and institutional issues cannot be resolved by simple rational arguments. It will take a greater fear, the fear of a real shortage of electricity, to get the country's attention relative to the need for a greater reliance on nuclear power. Electricity is almost as fundamental to our industrial society as air and water. A severe shortage can restrict industrial operations, cause massive unemployment, and impair food supplies. The inability to have an assured source to provide our homes and hospitals with lights, refrigeration, air conditioning, and heating, as well as radio and television, would have dire consequences. More extensive use of coal plants may be a short-term fix to an increasing demand for electricity. However, the possibility of irreversible environmental damage introduced by total reliance on coal may be a major incentive for the resurgence of U.S. nuclear power programs in the future.

2 Mother Nature

Mother Nature is usually personified as a kindly, benign friend. To some, our attempt to better ourselves is looked upon as a direct attack on nature that will lead to our ultimate destruction. However, our increasing presence on the earth and our quest for adequate food, clothing, shelter, and employment will necessarily cause significant changes in our surroundings. The earth now contains more than four billion people. By the year 2025, the number is expected to more than double to nine billion. A very large percentage of the current population lives in poverty characterized by starvation and disease. The projected population increase can only place further stress on the ecological balance. If our ultimate goal is to achieve a satisfactory quality of life for all the earth's inhabitants, our ability to feed, cloth, house, and gainfully employ such a huge number of human beings will be a massive undertaking that will affect all the earth's inhabitants and its resources. The situation will become more critical beyond the year 2025 and will further complicate our relationship with nature.

If we review our history on earth, it becomes obvious that we are really not very good friends with Mother Nature. In fact, we are normally adversaries. When the weather is cold, we must create our own heat. When the weather turns hot, we may prefer it to be cool. A soft rain waters the crops or the garden, but hurricanes and tornadoes can cause irreparable damage and death. Climatic changes causing droughts and deep freezes endanger our food supply. Storms and water runoff pollute our streams and erode our shore line. Earthquakes are always a threat to life and property.

The common assumption that "natural" is safe and "man-made" is suspect is contrary to current scientific knowledge. When the same criteria for analyzing cancer-causing substances in man-made food additives are applied to natural foods, it is found that essentially all foods in their natural state not only contain carcinogens but also possess other hazardous substances. It is hard to find any food that does not contain some harmful chemical that occurs naturally or is produced during cooking or by decomposition. Even a cup of coffee is estimated to contain more than 2,000 natural chemical components including caffeine. The caffeine in 100 cups of coffee is enough to cause convulsions in humans and possibly death if ingested in a single dose.

Plants are not simply foods for insects, birds, animals, and humans, but are highly complex organisms in their own right. The chemicals they synthesize defend the organisms against extinction. However, many of these chemicals in sufficient quantities can produce cancer in laboratory animals. The great bulk of our food, more than 99 percent by weight, consists of ingredients that nature put there, and man-made food additives make up less than one percent. Thus, human exposure to "chemicals" consists overwhelmingly of chemicals of natural, not man-made, origin.

To further illustrate the love-hate relationship between humans and Mother Nature, consider the sun. Without its rays there would be no life on earth, but ultraviolet radiation is the source of the most prevalent type of cancer, skin cancer, with about 300,000 cases reported each year. The sun is also the source of cosmic rays consisting of charged particles or ionizing radiation that represent a continuous exposure to all the earth's inhabitants. The earth contains vast amounts of minerals and chemicals necessary for survival, which include some materials such as arsenic and mercury that are very highly toxic and potassium, which is slightly radioactive. The earth also contains uranium and thorium, which remain radioactive for billions of years. Thanks to our ingenuity in developing the technology of converting uranium to heat-producing substances, we have the ability to generate a limitless supply of electricity should we choose.

The problem of how to live in harmony with Mother Nature is not an easy one. Various groups have sprung up in recent years whose primary goal is to protect the environment from the encroachment of humans. There have to be controls on arbitrary destruction of wilderness areas, wanton killing and threatening the existence of wildlife, and careless disposal of wastes. We want our air as clean as possible, and we must have an adequate supply of clean water. Most of us are against unnecessary pollution, so we can all be called environmentalists. Realistically, with an increasing population, it is difficult to improve our quality of life without expanding into unoccupied land areas, using renewable and nonrenewable resources where they exist, and endangering the lives of some plant and animal species. Because our actions are often irreversible, we must assure that current and often future activities will minimize the negative effect on the bounties nature has given us.

We do not know Mother Nature's plan for the earth. In the billions of years since the earth's creation, many changes have occurred. Furthermore, it is only in recent times, actually a few hundred years, that we have been able to learn and control some of nature's most complex secrets. We are finding that Mother Nature is far from perfect, and we can improve on her performance to our benefit.

In the course of the earth's development, many thousands of plant and animal species had disappeared well before humans could influence nature's actions. Even though some of our activities are causing other species to become extinct, some of these activities may be in our best interest, such as using pesticides to increase the yield of crops and satisfy the need for food. However, an endangered species law that protects all species whether or not they have significant environmental effect does not make sense. The "snail darter case" is a classic example of how interpretation of the law does not always serve the best interests of the public. The snail darter is a small, nearly extinct fish that resides in a river in Tennessee. The protection of this fish, whose demise should have no effect on the ecological balance, was considered to be more important than a hydroelectric plant that would produce clean electricity for thousands of people but would threaten the snail darter's existence. Carried to a similar extreme, we should consider initiating projects to resurrect species that are long gone on the supposition that humans know better than nature.

Mother Nature did not give humans a supplementary power source to make life easier. It was human ingenuity that led to the invention and application of steam and gasoline engines as well as the introduction of power plants to generate electricity. As a result, during the past two hundred years, humans have developed the technology to influence their own destiny, resulting in improving our quality of life and almost tripling our life expectancy.

3 A Brief History of Commercial Nuclear Power

Enrico Fermi demonstrated on December 2, 1942, that piling blocks of graphite that contained lumps of natural uranium resulted in a self-sustaining nuclear reaction. Although the power level was only a few watts, it was sufficient to show that uranium fission could generate heat in an easily controlled manner. Fermi's experiment is the basis for the nuclear power industry. All the subsequent research and development at government laboratories that applied this knowledge toward nuclear weapons has had no bearing on the development of commercial nuclear power plants. Protesting commercial nuclear power plants because of the bombings of Hiroshima and Nagasaki shows basic ignorance of nuclear power plant technology.

A nuclear reactor produces energy at a relatively slow rate, and under no condition can it give rise to an atomic explosion. A raging forest fire at the top of a mountain will produce a very large quantity of energy over a period of days, but there is no resulting explosion that can level the mountain. Violent explosions are produced by releasing a large amount of energy at a very rapid rate. Embedding TNT into the same mountain, which releases the same total energy as produced by the forest fire in a fraction of a second, will result in an explosion that can level the mountain.

Atomic bombs were being produced and detonated long before nuclear power came into existence. Commercial nuclear power plants have not contributed in any way to the materials or technology used in the manufacture of these bombs. The fact that we have chosen to develop the destructive potential of uranium so as to threaten our existence is a problem that would have to be resolved even if all the nuclear power plants on earth were retired from operation.

At the conclusion of World War II, Congress passed the Atomic Energy Act, which established the Atomic Energy Commission (AEC) and gave the AEC the responsibility for promoting and regulating nuclear power for peaceful purposes. The most obvious benefits of the new technology were the propulsion of submarines and the commercial production of electricity. Essentially all the initial AEC research and development funds were invested in projects that would demonstrate the feasibility of a reliable and safe nuclear reactor to propel U.S. Navy submarines. Because a nuclear power plant does not require air to operate, a submarine can stay submerged for months. Also, a nuclear power plant does not have to be refueled for years, so the support ships needed to fuel diesel submarines at sea could be eliminated.

Under the leadership of Admiral Hyman Rickover, the U.S. Navy undertook a massive scientific and engineering effort to develop nuclear propulsion power plants of the size and weight that would satisfy the Navy's submarine construction and operation requirements. With the launching of the nuclear submarine *Nautilus* in 1954, Rickover proved that nuclear power plants could be designed and constructed that pose no danger to the workers who fabricate nuclear power plant components and systems. More important, the Navy personnel who serve aboard these submarines live for months at a time in complete safety adjacent to nuclear reactors that produce the same type of radiation emanating from commercial nuclear reactors. (Figure 1 shows the interior layout of a U.S. nuclear submarine.) The size and health of the families of the hundreds of thousands of Navy personnel who have served aboard nuclear-propelled vessels are similar to those of the overall population. Navy-trained officers and enlisted personnel who have entered civilian life have assumed many key management and operation positions in the commercial nuclear power industry.

The success of the U.S. Navy nuclear reactor program opened the door to using this new technology to generate electricity commercially. A naval reactor produces much less energy than would be required for an economically viable nuclear power plant. However, the basic processes are similar. In the case of the naval reactor, the nuclear fuel is cooled by water in a pressurized vessel; this water is then converted to steam used to propel the vessel. In a commercial nuclear power plant, the nuclear fuel would also be cooled by water in a pressurized vessel. The water would be converted to steam to drive a turbine-generator and produce electricity. There are many design differences because a naval reactor has to satisfy stringent weight and size considerations and also respond to special tactical

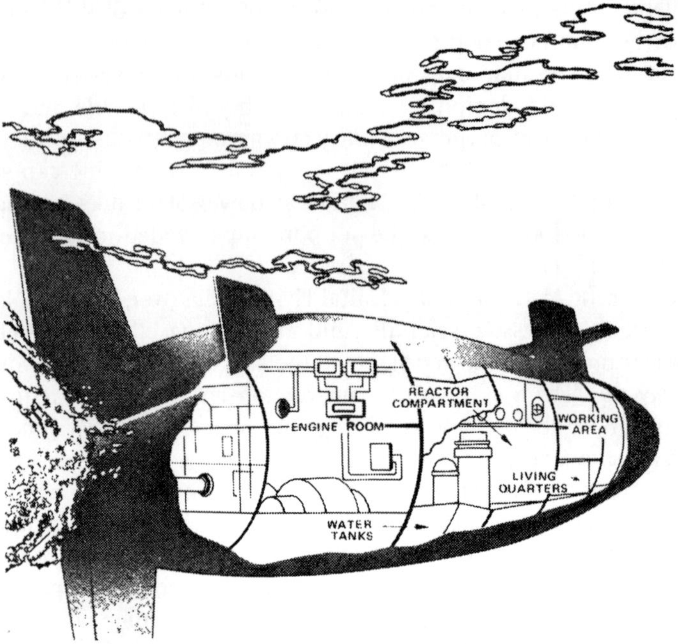

For more than thirty years, the U.S. Navy has used nuclear power plants to propel submarines and surface ships. The internal structural design of these vessels protects all personnel from excess radiation from the nuclear reactor. Hundreds of thousands of Navy personnel have served aboard these submarines and ships in complete safety.

The design of the nuclear propulsion system permits these vessels to go from rest to full speed in a short time. The nuclear reactors used in commercial nuclear power plants are similar to the nuclear reactors used by the Navy. The Navy model may be considered similar to an engine that drives a small car, while the utility model may be considered similar to a very large engine that drives a locomotive that hauls freight trains. The technology developed for the U.S. Navy nuclear propulsion program has formed the basis for the commercial nuclear power plant program. Navy-trained officers and enlisted personnel experienced in the operation and maintenance of naval reactors are now a vital part of the commercial nuclear power industry.

Figure 1. A typical interior layout of a U.S. nuclear submarine

situations, such as maneuvering from low to high speed in a relatively short time. Another significant difference involves the reactor fuel itself. For naval applications, the fuel must have a special configuration and composition that is much more expensive than the fuel needed for commercial application.

Using the technology developed to propel submarines, the U.S. Navy nuclear reactor program was able to design and construct the Shippingport Atomic Power Station, the nation's first large-scale commercial nuclear power plant. This project was started in 1953 to confirm the practicality and safety of nuclear power for commercial application. The plant was designed initially to generate 60 megawatts of electricity and was commissioned to operate in 1957. It was built near Pittsburgh, Pennsylvania, and operated by the Duquesne Light Company. Shippingport did demonstrate the feasibility and safety of this type of plant. After more than 20 years of operation, it was recently retired from service and is currently being decommissioned.

4 Why Nuclear Power?

With all the difficulty in gaining public acceptance today and the economic problems that have been encountered in the nuclear power program, it is natural to wonder what the bases were for the initial decisions to attempt to generate a major portion of the nation's electricity by nuclear power.

If we go back to the early 1960s, the decisions that led American utilities to pursue the nuclear option made a lot of sense. Billions of dollars had been spent in research and development in the 1950s to develop safe and efficient nuclear reactors to propel U.S. Navy submarines. The success of the U.S. Navy's nuclear program and the Shippingport Atomic Power Station showed that the technology had been developed to the point that it should be relatively straightforward to build nuclear power plants of sufficient size to produce electricity and be economically competitive with fossil plants. The technology still had to be tested by constructing and operating several different types of small prototype commercial nuclear plants to evaluate performance, safety, and economics.

As a result of the technical evaluation of the performance and safety of these prototype plants, industrial organizations in the United States decided that water-cooled and -moderated reactors of the pressurized or boiling type were best suited for commercial application. However, a major concern was whether a nuclear power plant would be economically competitive with fossil plants. When detailed economic studies indicated that large nuclear plants could be built and operated at costs less than fossil plants, a new industry was born. Companies such as General Electric and Westinghouse guaranteed that the costs of the first commercial nuclear power plants sold to the utilities would not exceed those of conventional fossil plants. These plants performed well enough to convince a large number of utilities that nuclear power plants were both safe and economically feasible.

From the experience of building and operating fossil plants, it was believed that to achieve maximum economic benefit from nuclear power plants, meaning smaller monthly bills to the rate payers, these plants should be as large as practical. By this economy of scale, the increased incremental costs associated with design, construction, operation, and maintenance of a very large plant would be more than offset by lower total capital, operations, and maintenance costs per unit of electrical production. For example, one 1,000-megawatt plant should be less expensive to build and operate than two 500-megawatt plants.

Proponents of nuclear power were so enthusiastic about the economic benefits that would be derived from very large nuclear power plants that some were even quoted as saying that electricity from nuclear plants would be "too cheap to meter." Obviously this was an oversimplification, as there had to be some charge, no matter how small. Also, before the 1960s, as a result of higher thermal efficiencies, reduced capital costs per installed kilowatt of new, large fossil plants, and the low price of oil, the cost of electricity to the rate payer had been steadily decreasing, and it appeared that nuclear power plants would continue this trend.

In the early 1960s, the government and industry did not anticipate any unusual technical or economic problems associated with nuclear power plant design, construction, and operation. No restrictions were placed on the number of power plants that could be built. Most of the larger utilities and even some of the smaller ones were totally convinced that nuclear power would best satisfy their customers. In the late 1960s and early 1970s, the orders for nuclear plants began flooding in—46 were ordered in 1973 alone. At that time, these new plants seemed priced reasonably enough, estimated to cost in the neighborhood of $200 million to $300 million. As explained by the manufacturers, a power plant was a power plant, and nuclear fission was just another way of heating water.

In fact, the processes used by fossil plants and nuclear plants to produce electricity are quite similar. In the first case (see Figure 2), the combustion of fossil fuel with air produces heat, which is ultimately converted to the steam that turns the turbine and generates electricity. In the second case (see Figure 3), the fission process within the nuclear fuel produces the heat that goes through a similar conversion process to generate electricity. However, because of

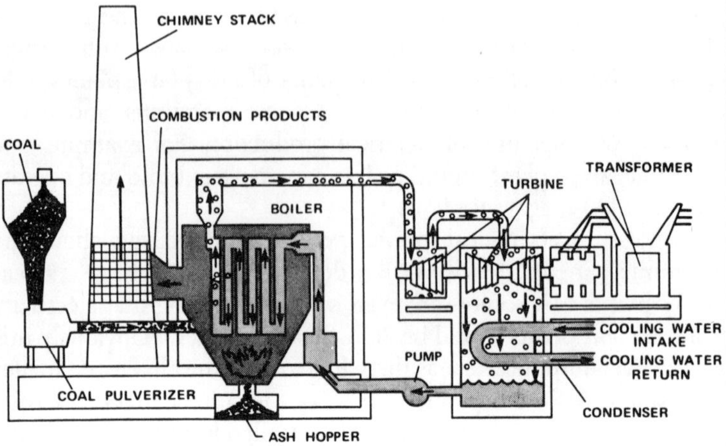

Coal from the mine is pulverized and burned in the boiler. Water circulating through the boiler absorbs the heat and changes to steam. The pressure of the steam against the turbine blades spins the shaft and turns the generator. This movement of the generator produces electricity, which is modified by a transformer and sent by transmission lines to be used by homes and industry. Used steam cooled by the condenser is changed back into water for reuse in the boiler. The cool water that condenses the steam is drawn from a river, lake, or ocean. If it is necessary to satisfy environmental requirements on the temperature of the water discharged back to the source, auxiliary cooling systems, such as cooling towers, are employed.

The combustion products resulting from the burning of the coal, which include sulfur dioxide, nitric oxide, and carbon dioxide, are discharged into the air through the chimney or stack. The accumulated ash, which is a solid waste, must be removed for land burial.

Figure 2. Electricity produced by coal

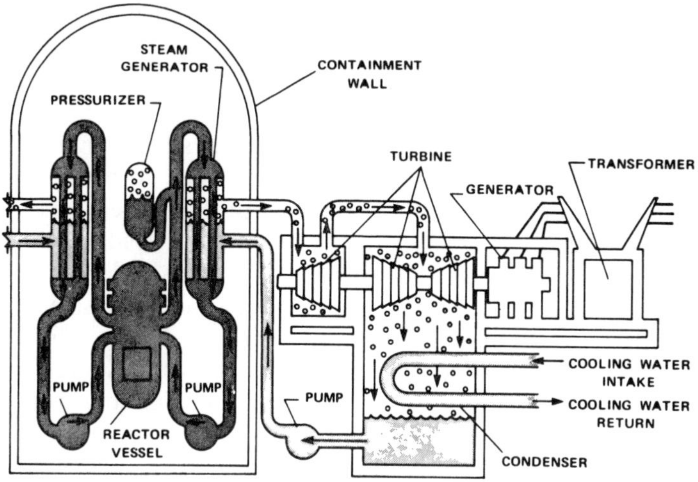

Nuclear fuel produces heat by controlled fission in specially fabricated cylindrical fuel rods located in the reactor vessel. In a pressurized water reactor, the water under pressure circulates through the reactor vessel and is heated by the fuel rods. This heated water leaves the reactor vessel and enters a steam generator. Here the water from the reactor gives up the heat to a separate water system, turning the water into steam, and the water originating in the reactor is returned to the reactor at a lower temperature. The steam from the steam generator turns the turbine-generator. Used steam cooled by the condenser is changed back into water for reuse in the steam generator. The electricity produced is modified by a transformer and sent by transmission lines to be used by home and industry. Auxiliary cooling systems such as cooling towers are used to assure that there is little or no increase in cooling water discharged back to its source.

Figure 3. Electricity produced by nuclear fuel

possible safety hazards associated with the leakage of radioactive materials from the system, the materials and construction of nuclear plants, including quality assurance and quality control programs, are much more intricate than those associated with fossil plants. This fact did not receive the attention it deserved in the early days of the nuclear power program, and the industry has yet to recover from this oversight.

The early sales success of the manufacturers has proved to be a liability. Any first-of-a-kind plant is very likely to have design, construction, operation, and maintenance problems. Because of the large number of initial sales, there was not time to identify these problems and eliminate them from future plants. Oddly enough, most of the problems requiring major modifications and significant cost increases are not attributed to nuclear technology. Instead, weaknesses in conventional construction and the malfunctioning or metallurgical defects associated with such conventional components as pipes, pumps, and valves have proved to be the most troublesome.

Despite technical difficulties, cost overruns, and the Three Mile Island accident, the U.S. nuclear industry has been able to demonstrate that the technology works. In 1986, there were 102 nuclear plants licensed for commercial operation in the United States generating about 17 percent of the nation's electricity with ten additional plants in the startup phase and another thirteen in the construction phase. When and if all these plants go into operation over the next few years, nuclear plants will account for nearly 20 percent of the nation's electricity. Since the United States generates about one-third of all the electricity in the world today, this figure represents an enormous amount of electricity. During the Arab oil embargo in 1973, oil generated 17 percent of the nation's electricity and nuclear power generated 5 percent. In 1986, increased use of nuclear power has interchanged these numbers with nuclear power at 17 percent and oil at 5 percent.

In this book, the words "nuclear power plant" and "nuclear unit" are used interchangeably. Ninety-one sites accommodate the 125 nuclear power plants currently scheduled for operation. Thus, several sites have more than one nuclear unit. Some of these sites can accommodate as many as six units. Currently, the maximum number of units per site is three. Figure 4 shows a typical arrangement of a two-unit nuclear power plant. Figure 5 shows the nuclear power plant sites in the United States.

The architectural design of a nuclear power plant is usually very attractive, blending into the local surroundings. The appearance of the plant both from the inside and outside can be closely compared to a medical laboratory rather than an industrial plant. The large cylindrical structures are the containment buildings that house the nuclear reactors. These structures are the extra safety margin that guarantees that, in the unlikely event large sources of radiation were to escape from the reactor, the radiation sources would be retained in the containment building and not be discharged into the atmosphere. The total reactor site, including all the auxiliary electrical and cooling equipment, occupies 500 to 1,500 acres.

Figure 4. Typical arrangement of a two-unit nuclear power plant

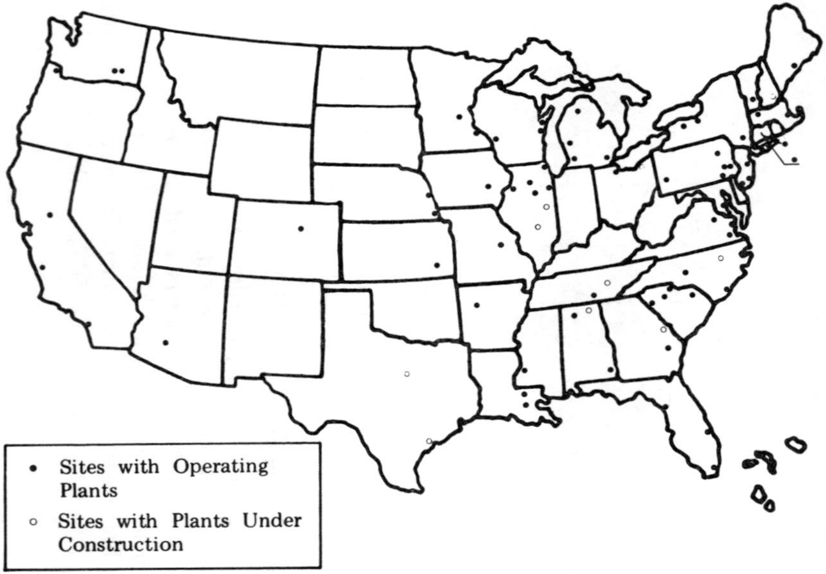

Today more than 100 commercial nuclear power plants are licensed to operate, and approximately 25 additional plants are scheduled to go into operation. Only about one-third of the heat generated by these plants can be converted into electricity. The remaining heat must be removed from the plant by cooling water. For this reason, plant sites are usually located adjacent to large sources of cooling water such as rivers, lakes, and oceans. The sites selected must satisfy stringent federal and state regulations governing water quality and temperature, earthquakes, and release of radioactivity. Some sites can accomodate up to six nuclear units, but usually a site contains one or two units.

Figure 5. Nuclear power plant sites in the United States

It may be surprising to know that more than 250 nuclear plants are now in operation outside the United States. Moreover, 136 plants are under construction and an additional 172 are in the planning stage. Most of these plants use the water reactor technology developed in the United States. They also use American codes and standards in the design phase, and many American companies are involved in the construction and operation phases. More than half of the electricity of France and Sweden is generated by nuclear power plants. Other countries, such as Japan, West Germany, the United Kingdom, and the Soviet Union, are more dependent on nuclear power than the United States. All these countries are committed to build nuclear plants to satisfy their future electricity requirements.

The accident at the Chernobyl nuclear power plant in the Soviet Union (discussed in Chapter 8) is a severe setback to the Soviet's nuclear program. This accident will result in major modifications to Soviet reactor designs and plant safety systems as well as an increase in the level of training of all Soviet plant operators. Nevertheless, despite the accident, the Soviet Union plans to expand its nuclear power program. Even though the Chernobyl accident is unique to this particular reactor design, the magnitude of the accident raises questions as to the safety of all nuclear power plants and probably will result in deferment or cancellation of some plants that were in the planning stage.

5 Regulation of Nuclear Power Plants

The original Atomic Energy Act in 1946 that created the Atomic Energy Commission (AEC) gave the AEC the charter to both promote and regulate the use of atomic energy. With the growth of the nuclear industry, it became necessary to separate these two functions, and the AEC was phased out of existence in 1974. The U.S. Nuclear Regulatory Commission (NRC) was given responsibility for the licensing, regulation, and safety of commercial nuclear power plants.

The transition from the AEC to the NRC served to politicize the nuclear power program in the United States. Before the advent of the NRC, many proponents of nuclear power in the government, including Congress, encouraged and supported this new technology. After the formation of the NRC, the political climate changed. Criticism of the nuclear power program by a small but very vocal opposition made it politically expedient to be against nuclear power and politically hazardous to express public support for nuclear power.

The NRC is governed by a five-person board headed by a chairperson, each nominated by the President and approved by the Senate for a five-year term. In 1977, President Carter nominated as commissioners two persons who were against nuclear power. These appointments resulted in a weak and divided Commission that could not or would not make decisions and further eroded the confidence of the American public. This situation is comparable to a local public school board that, in order to represent all viewpoints, has members who are opposed to public schools and in favor of private schools. The actions of such a board relative to administering and improving the public school system would be a continued source of controversy.

President Carter's energy policy also undermined the nation's confidence in nuclear power. He totally surrendered to the nuclear opposition when he stated that nuclear power was the electrical energy source of last resort. His order of priority was coal, solar and, if necessary, nuclear. His viewpoints were translated into government action or inaction and were a major blow from which the industry has not yet recovered.

The total staff of the NRC is about 1,500, of which 110 are full-time legal staff. The United Kingdom's Sir William Marshall, Chairman of the Central Electricity Generating Board, has noted that the United Kingdom, which has 36 commercial nuclear plants in operation and 7 under construction, has a total staff of 97 and no full-time legal staff. He deduced from the comparison that the United Kingdom's licensing system depends entirely on the judgment of science and engineering and not on the interpretation of words. A safety document in the United States will total nearly 3,000 pages when a comparable document in the United Kingdom is 45 pages. The fact that the United Kingdom has achieved an enviable safety record in operating its commercial nuclear power plants indicates that more regulation does not mean better safety.

The Presidential Commission that investigated the Three Mile Island accident was highly critical of the NRC. It noted that NRC regulations are too voluminous and too complex, requiring immense efforts by the utilities and their suppliers to assure compliance. It concluded that the NRC organization at that time lacked the organizational management necessary for the effective pursuit of safety goals. It recommended that the NRC be abolished and replaced with an executive agency, headed by a single administrator, with the responsibility of ensuring public safety as its primary mission.

The NRC's reaction to the criticism resulting from Three Mile Island has been to generate hundreds of new safety regulations. For those plants still under construction, the utilities had no choice but to tear apart nearly completed plants or redesign plants just under way. The NRC paid little attention to the cost effect of the regulations it imposed. The industry itself was on the ropes, and most companies were prepared to do whatever was asked of them, regardless of the cost.

Many of the new NRC regulations have been imposed arbitrarily. In some cases, correcting a single problem could make the plant less safe because of the solution's effect on other aspects of operation. As an example, the publicity associated with potential terrorist attack on nuclear plants has led to requirements for very sophisticated security systems that limit access of all personnel to various parts of the plant. While these procedures limit outsiders from entering

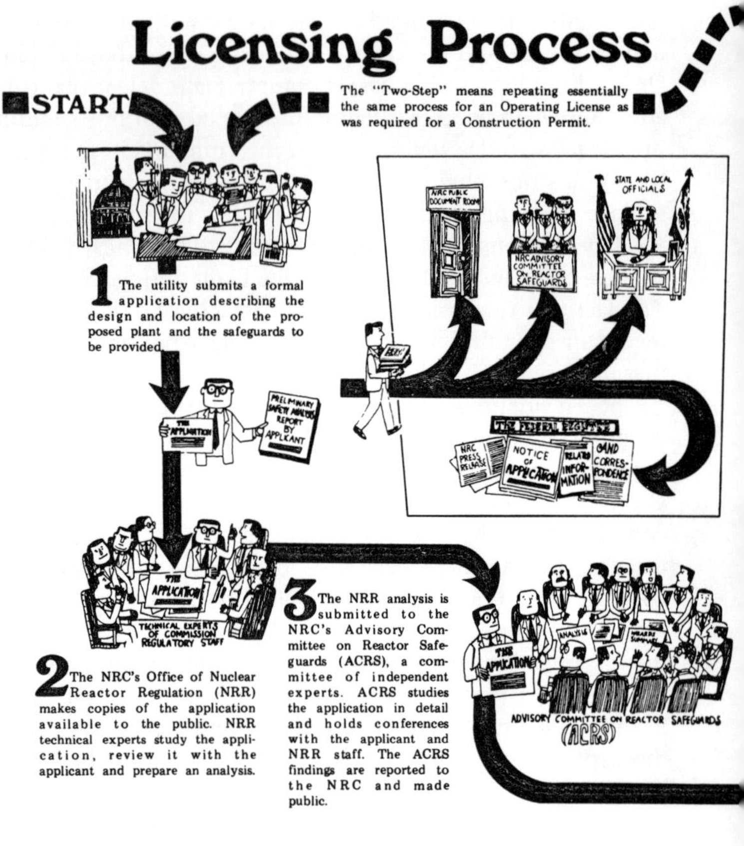

The licensing process requires two separate licenses, one to construct the plant and the other to operate it. It has been taking about 6 years to complete all of the requirements to obtain the construction permit. About another 5 years is required to complete construction and obtain an operating license. Finally, about one year of testing is required before the plant is placed

Figure 6. The NRC licensing process

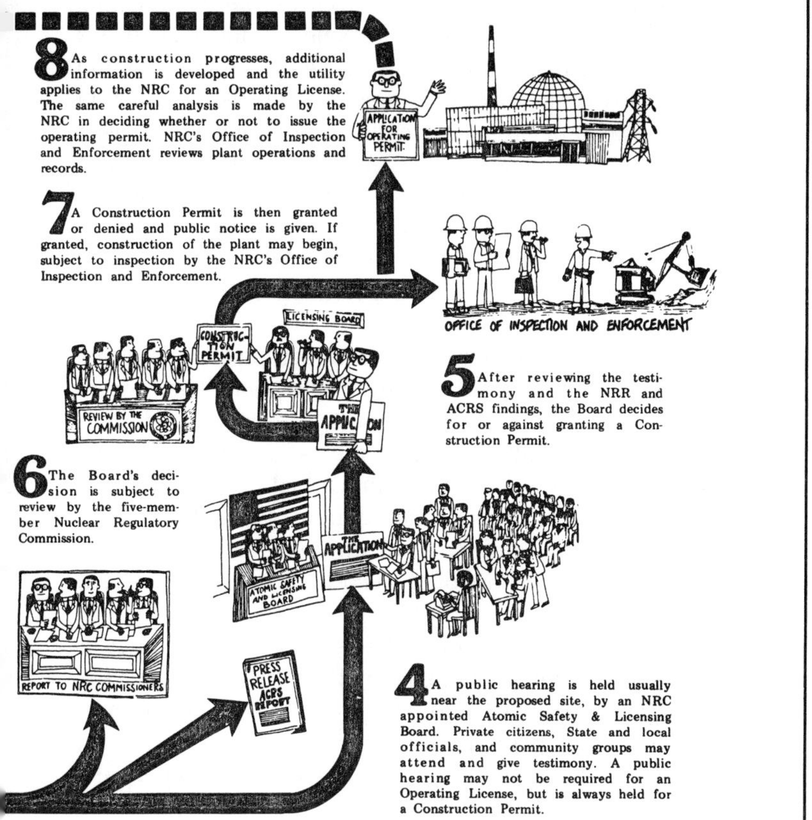

into commercial operation. All during construction and even after operation, the plant is under inspection by the NRC to make sure the plant is in compliance with federal regulations. In addition, the utility must report any abnormal occurrences or face severe penalties including loss of license.

the plant, they also make it difficult for plant workers to carry out such normal functions as equipment maintenance.

The current NRC licensing process, shown in Figure 6, is a two-step process that requires two separate licenses, one to construct the plant and the other to operate it. It generally takes about six years to complete all the requirements to obtain the construction permit. Another five or six years is required to complete construction and obtain an operating license. Finally, approximately one year of testing is required before the plant is placed into commercial operation. All during construction and even after operation, the NRC periodically inspects the plant to make sure it is in compliance with federal regulations. In addition, the utility must report any abnormal occurrences or face severe penalties including loss of license.

This NRC licensing process would not be acceptable to a utility that is considering building or ordering a new plant. The process is too cumbersome, too lengthy, and strewn with pitfalls. There is no guarantee that, after all the requirements are satisfied, some arbitrary court action by a disgruntled individual or group will not delay plant operations for an inordinate amount of time. Since delay in obtaining an operating license results in increased finance and labor costs and is a major cause of huge cost overruns, this concern is very valid.

With all the experience accumulated here and overseas in nuclear plant design, construction, and operation, the time to complete the process in the United States could be cut in half without compromising the safety of the plant. Furthermore, the process itself must become one step; that is, if the utility applicant satisfies all the requirements for construction and operation, it should be permitted to construct and operate without further qualifications.

6 The Three Mile Island Accident

On March 28, 1979, Three Mile Island (TMI) Nuclear Station, Unit 2, experienced the worst accident in the history of commercial nuclear power operations in the United States. As a result of human, institutional, and mechanical failures, the inventory of water that cooled the nuclear fuel assemblies was allowed to decrease to the point where the fuel was no longer submerged in water. As a result, the metallic cladding that surrounded the fuel and some of the fuel itself melted, releasing large quantities of radioactive materials into the leaktight containment building.

With all the care and precaution that go into the design of a nuclear plant, how could such an accident have happened? Nuclear plants are designed with redundant safety systems and built-in operating limits. By preventing the plant operators from exceeding these limits, the plant protection systems mitigate against human error. On the other hand, should either plant equipment or plant systems malfunction, the plant operator has the necessary skills to shut the plant down manually and maintain it in a safe condition. The operators can also keep the plant on line when faced with malfunctions of non-safety-related equipment. Avoiding unnecessary shutdowns can save utility customers about $500,000 to $1,000,000 per day depending on the source of replacement power. This balance between operators and automatic plant systems minimizes the potential for accidents while obtaining the maximum economic benefits.

At TMI, the system broke down because the four licensed and qualified operators misinterpreted the chain of events and overrode safety systems that were installed to handle such an event. They interpreted the sequence of events to be an overpressurization of the primary coolant system, meaning an excess of water, which could rupture the system. To protect against this condition, the operators consciously turned off emergency systems that would have added cooling water to the fuel so that it would never have become uncovered. This situation is somewhat analogous to the actions of a crew flying a plane in heavy fog at night through a mountainous area. The crew experiences difficulties with the proper operation of several systems. They become convinced that an altimeter instrument reading of 20,000 feet is correct even though the behavior of the instrument and other indicators should have alerted the crew that the altimeter reading is wrong. They keep the plane at ground level when in fact they should be flying above the mountains and they finally crash into a mountainside.

To gain some insight into the operators' decisions that led to the TMI accident, it is necessary to refer to the plant design shown in Figure 3. A major component in the reactor coolant system is the pressurizer that figured prominently in the TMI accident. The pressurizer controls the reactor coolant system pressure and water level. It is partly filled with water, and its upper part contains steam. If the pressure drops, heaters in the pressurizer are turned on to raise the water temperature and form steam, which raises the system pressure. If the pressure rises, a spray valve containing cold water is turned on, which condenses some of the steam and reduces the pressure. The pressurizer also has at its top an electromagnetic relief valve that can be opened if pressure rapidly increases.

At about 4:00 a.m. on March 28, 1979, a series of minor malfunctions occurred that resulted in no water entering the steam generators and no steam leaving. The reactor coolant temperature rose, causing the coolant to expand and creating increased pressure. To relieve this pressure, the electromagnetic relief valve on the pressurizer opened so that enough coolant would escape to return the system pressure to normal. Because of the increased system pressure, the reactor automatically shut down while coolant was released through the open relief valve. It was about 13 seconds before the reactor coolant pressure returned to normal and a light appeared on the control room panel indicating that the valve had closed. The light, however, showed only that a pressure demand signal had been given to the valve; there was no indication to show whether the valve was actually open or closed. In fact, it was stuck open. For more than two hours, a serious loss of coolant was occurring, eventually uncovering the fuel in the reactor vessel, leading to some fuel melting and radioactivity escaping into the containment building.

The operators could have closed the stuck-open relief valve manually and totally prevented any serious damage. Although the spurious signal indicated that the relief valve was closed, there were other indications that it was open, which were ignored by the operators. In addition, every nuclear reactor is designed to prevent a potential loss of reactor coolant. As the water level and system pressure dropped because of the open relief valve, it eventually reached a setpoint at which the emergency core coolant system would automatically inject water into the pressure vessel. This system worked as designed. The pumps that activate this emergency system began delivering water to the coolant system. The coolant pressure began to rise again. Had these pumps been left to do their designed task, the accident at TMI could have been averted. After about four minutes of emergency system operation, the operators made their major error in judgment. They began to throttle back on one of the emergency pumps and turned the other one off completely.

Why did the operators throttle back an emergency system that was fulfilling the very role for which it was designed? The answer is complicated. When the relief valve opened, the steam in the pressurizer was the first to escape, dropping the pressure in the reactor vessel and causing a steam bubble to form under the vessel head and water to be displaced from the vessel into the pressurizer. This caused the pressurizer level indication to rise, until at about 6 minutes into the accident the level went off scale, indicating the pressurizer was full of water.

The operators at TMI had been trained to avoid a solid pressurizer (meaning a pressurizer totally filled with water) by cutting back on injection flow. Pumping water into a system with a solid pressurizer could exceed design pressure and rupture the system. What the TMI operators did not realize was that the system was by no means filled with coolant. While the coolant level in the pressurizer was extremely high, the coolant had become a mixture of steam and water, and the amount of water was rapidly decreasing.

Another operator error involved shutting off two of the four reactor coolant pumps at 74 minutes into the accident. At 101 minutes they shut off the remaining two pumps. Why did the operators turn off the reactor coolant pumps? Because of the low pressure and coolant flow, high vibrations coming from the pumps could be felt in the control room. It is an industry procedure to protect the pumps from potential damage unless there is a nuclear accident involving a loss of coolant. Unfortunately, the TMI operators had not determined that they were experiencing a loss-of-coolant accident. About two hours into the accident, the fuel had been sufficiently uncovered as to reach the point of no return. By that time, the cladding of the fuel rods was deteriorating from oxidation by steam, and radioactive fission products were escaping from the fuel to be swept away by the steam and water mixture into the containment building atmosphere.

Essentially all the radioactive material was retained in the containment building and did not escape outside the building. The drain tank below the pressurizer did overflow because of the leakage from the stuck-open valve during the early stages of the accident. The water contained low-level radioactive materials. A human error was involved in permitting a large volume of the radioactive water to be pumped from the containment building, which is leaktight, to an adjacent auxiliary building which is not leaktight. However, the total release of radioactive materials was not "significant," resulting in a small temporary increase of radiation levels beyond the plant site boundaries.

"Significant" is a relative term. Had there been a major radioactivity release, people living in the vicinity of the plant could have been overexposed. As it turned out, the maximum radiation exposure to a few individuals living near the plant could have been as much as the equivalent of four chest X rays. The various commissions and investigative groups established by the federal government and the State of Pennsylvania concluded that, since the total amount of radioactivity released was so small, no individual could have experienced any adverse health effects as a result of the TMI accident. They also stated that there will be no detectable birth defects resulting from the very low radiation exposure of the general public and pregnant women. Within months of the accident, the Pennsylvania Department of Health initiated several studies to evaluate the possible long-term health effects of the accident on 35,000 people who lived in the immediate area of the plant. Of particular concern were pregnancy outcomes, infant health, and cancer. In 1986, seven years after the TMI accident, the findings of the Department of Health show that there is no evidence of any increased health risk to the new born or the living as a result of the accident.

7 Protection against the Unexpected

As an aftermath of the Three Mile Island accident, the obvious question that the public would want answered is "Are nuclear plants safe?" or phrased in another way, "Are nuclear plants safe enough?" To answer these questions, we must better understand how the public looks at safety and how it reacts to an accident. When the space shuttle Challenger blew up, killing six astronauts and one schoolteacher, it was a total shock to the entire nation. Launches for manned space missions had become almost routine. We had come to expect that all such missions would be successful. A few months before the shuttle accident, nearly 250 American paratroopers were killed in a plane crash returning from a tour of duty in the Middle East. This tragedy did not receive nearly the same attention as the destruction of Challenger because we have come to accept a plane crash, no matter how great the loss of life, as a normal occurrence.

We were not psychologically prepared to expect an accident in the space shuttle program, particularly one that involved a civilian schoolteacher. The number of people killed was not as important as the fact that we had become convinced that accidents of the Challenger type could not occur. In fact, if we look at the huge quantities of combustible materials required for a shuttle launch, the program's safety record has been remarkable. We should all have been prepared for the fact that an occasional failure involving loss of life would occur. Investigation into the Challenger destruction revealed that an accident was inevitable because of design deficiencies in the solid booster rocket. This deficiency is being corrected so that the probability of failure of future manned space launches should be greatly reduced.

In the discussion of safety, nuclear power plants should not be compared to airplanes or the space shuttle. A malfunction that can lead to a serious accident on a plane or space shuttle can occur in seconds or less. With all the malfunctions and human errors that occurred at Three Mile Island, it took nearly two hours before fuel damage occurred. No one has ever said that accidents would not happen as a result of nuclear plant operation. Like any other large industrial plant, nuclear plants have accidents or at times exceed limits on radioactive discharges from the plant. In all nuclear plants that have operated, personnel have erred, electrical systems have failed, pumps have failed, valves have not opened or closed as they should, and radioactive discharges from the plant have temporarily exceeded license requirements. Some of these accidents or effluent releases have required relatively minor repairs or changes in operating procedures and may not have involved a plant shutdown. In other cases, equipment damage may have been severe, resulting in a lengthy and costly shutdown. However, all accidents that occurred in commercial nuclear power plants before Three Mile Island posed little or no threat to the public because very little radioactivity was released within the plant.

The engineers and scientists who design nuclear power plants recognized that it is impossible to include in the design analysis all possible malfunctions and abnormal occurrences that may lead to an accident that releases large quantities of radioactive materials. Therefore, the designers insisted that in the unlikely event of a major accident caused by multiple failures, an outer structure or containment building must be an integral part of the power plant. The sole purpose of the containment building is to contain all the radioactivity that could result from the worst possible accident, so that the people living in the surrounding area would not be affected. Thus, to protect against some unanticipated event or events not considered in the plant design and causing release of large quantities of radioactivity within the plant, every U.S. commercial nuclear plant has an outer structure or containment building that prevents radioactivity from escaping into the atmosphere. The containment building protects the public against the unexpected such as occurred at Three Mile Island.

The containment building shown in Figure 7 does not look impressive from the outside because the lower half is usually underground. It is overshadowed by the cooling tower, described in Chapter 21, which is 500 feet above ground. A typical containment building for a pressurized water reactor is cylindrical in shape, about 300 feet high and 130 feet in diameter, with reinforced concrete walls nearly 5 feet thick and a dome about 3 feet thick. The building

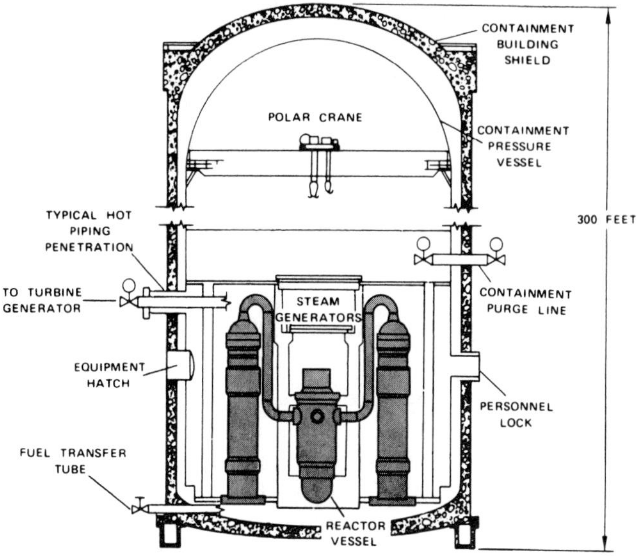

To protect against the possibility of a catastrophic accident that would endanger public safety, all nuclear plant systems that would release radioactivity in an abnormal occurrence are housed in a specially designed containment building. In the unlikely event that the worst possible accident were to occur and the reactor were seriously damaged, no radioactivity would escape from the containment building, and the public would not experience any adverse health effects.

A typical containment building for a pressurized water reactor is cylindrical in shape, about 300 feet high and 130 feet in diameter with reinforced concrete walls nearly 5 feet thick and a dome about 3 feet thick. The building is sealed at all times with entry and exit permitted only through specially designed chambers, and all penetrations have containment isolation valves that close automatically under accident conditions.

Figure 7. A typical containment building

is sealed at all times, with entry and exit permitted only through specially designed chambers, and all penetrations have containment isolation valves that close automatically under accident conditions.

The analysis of the Three Mile Island accident revealed that under the worst conditions, the containment building would not have been breached and there was no possibility of a release of hazardous amounts of radioactive material from the plant. In other words, the public was never in any real danger. However, the uncertainties that existed at the time warranted the alerting of nearly 600,000 people for possible evacuation. The containment building behaved as designed. Although some of the fuel overheated and melted and released radioactive materials, all this radioactivity remained within the containment building. The destruction of the reactor was an economic disaster but in no way affected the health of the populace surrounding the plant.

The next question that needs to be answered is: "What is the probability of an accident that also ruptures the containment building, thereby subjecting the surrounding population to radiation levels well above normal?" Attempts have been made to answer this question by developing possible scenarios and assigning probabilities to the occurrence of such an event. In 1973, the Atomic Energy Commission set up a panel of 60 scientists, headed by Professor Norman Rasmussen of the Massachusetts Institute of Technology, to perform a risk assessment of nuclear power plants. The Rasmussen Report, published two years later, stated that the risks of a catastrophic event were negligible. In fact, it concluded that meteors from outer space are a greater threat, statistically, than major nuclear accidents. These findings were endorsed by the NRC as an accurate measure of the safety of nuclear plants.

In 1977, the NRC formed a Risk Assessment Review Group, headed by Professor Harold Lewis of the University of California at Santa Barbara, to independently evaluate the Rasmussen results. The Review Group contended that the Rasmussen Report understated the numerical estimate of the overall risk of a reactor accident. The Review Group made no attempt to quantify the higher probabilities, so it is not known by what factors the conclusions of the Rasmussen Report would be changed. As a result of the findings of the Risk Assessment Review Group, the NRC withdrew its endorsement of the Rasmussen Report. Therefore, we now have no official government position on the probability and consequences of a catastrophic event.

Without experimental evidence, it is not uncommon for scientists and engineers to disagree on the theoretical prediction of an event that has never occurred. No numerical values assigned to nuclear safety can be final; they depend on subjective evaluation of the current state of technology. The only proof of risk assessment is experience, and the experience is that there are no known fatalities associated with nuclear plant operation in the United States. We would all like to be told that the probability of an accident resulting in a large release of radioactivity to the atmosphere is zero or it can never happen. The best that can be said is that the probability is very small or close to zero.

Natural and man-made disasters are probably more of a threat to the well-being of many of our citizens than nuclear power plant accidents. An earthquake in a populated area in the state of California, collapse of a large dam, and a jumbo jet with a full complement of passengers crashing on takeoff or landing into a downtown stadium containing 50,000 people are events that could occur in our lifetime. Massive flooding of low-lying coastal areas due to hurricanes, tornadoes traversing populated areas, explosions at industrial sites producing toxic chemicals, and transportation accidents involving potential lethal cargoes are life-threatening events that occur relatively frequently in the United States. People other than those directly affected tend to consider these events of local interest.

The occurrence at the Three Mile Island plant is best described as a small loss-of-coolant accident based on multiple failures. The Rasmussen Report devoted relatively little attention to this type of accident, having focused much of its attention on sudden rupturing of large coolant pipes and the subsequent total loss of primary coolant. The nuclear plant safety systems in use before the TMI accident and those installed as a result of lessons learned from the TMI accident are designed and operated to guarantee that coolant water will always cover the fuel. Obviously, the best solution to a severe accident is not to have any. The primary objective of the reactor safety design is to assure that, if an unforeseen accident occurs, no damage is done to the fuel and therefore no radioactivity can be released. Changes in operational procedures, modification of existing safety systems, and incorporation of new safety systems are all intended to prevent damage to the fuel under any accident conditions. If, despite all these precautions, an accident occurs resulting in the melting of fuel accompanied by radioactive release, all the radiation will be confined in the containment building and the public will not be affected.

8 The Chernobyl Accident

Not all nuclear power plants are the same. The Soviet Union has a unique reactor design and safety philosophy that is significantly different from nuclear plants designed, built, and operated in the United States. There are similarities between the two designs. Both types of reactors use fuel assemblies composed of many rods, each containing uranium dioxide clad with zirconium alloy. Both types of reactors use water to cool the fuel assemblies and produce steam to drive a turbine-generator to generate electricity. In the Soviet design, the fuel assemblies reside in individual pressurized tubes or channels composed of zirconium alloy and embedded in graphite. A typical Soviet reactor has approximately 1,660 such channels and contains 1,700 tons of graphite. In U.S. commercial reactors, all the fuel assemblies reside within one very large pressure vessel that does not contain graphite.

There is one major difference in the normal operational characteristics of U.S. and Soviet reactors. In a U.S. reactor, an increase in power produces a decrease in water density, which reduces the fission rate and causes the power output to level off. In this manner, the power output is self-regulating and the reactor is inherently stable. In a Soviet reactor, an increase in power also produces a decrease in water density, but the decrease results in an increase in the power, so that mechanical controls must be inserted to reduce the fission rate in order for the power output to level off. In the Soviet design, the power output is not self-regulating, and the reactor is unstable and more difficult to control, particularly under extraordinary circumstances.

Both U.S. and Soviet nuclear plants have emergency cooling systems that are actuated automatically when the amount of water cooling the fuel is reduced below a specific volume. Both types of plants have control systems that will automatically shut down the reactor if specific temperature, pressure, and flow conditions are not satisfied. All U.S. commercial plants have containment buildings (as described in the previous chapter) to prevent the venting of radioactive materials directly to the atmosphere in the event of an unforeseen occurrence. The Soviets have stated that their plants were so safely designed that to have a containment building was not necessary. Another consideration in the Soviet decision not to have a containment building is that their nuclear reactors occupy a much larger volume than U.S. reactors and require a much larger structure, adding significantly to the complexity and the cost of the plant.

On April 26, 1986, reactor operators were conducting a safety test on Unit 4 at Chernobyl Nuclear Power Plant. The purpose of the test was to determine how long the steam-driven turbines at the plant would continue to generate electricity in the event of an electrical blackout at the plant. Normally, backup diesel generators would start up automatically to keep the plant systems in operation. It appears that the Soviet's diesel generators have been slow to start up. The objective of the safety test was to determine the length of time the turbines could supply power without relying on the diesel generators.

In the course of carrying out the safety test, the reactor operators violated six key safety procedures. In their haste to complete the test they deliberately shut down, disengaged, and overrode safety systems that were specifically designed to automatically shut the plant down under abnormal conditions. These actions resulted in a large increase in the reactor power level, which caused the water to boil and decreased its density. Because of the inherent instability of the reactor, the power output continued to surge, resulting in almost total evaporation of the water cooling the fuel and the formation of very large quantities of steam. The steam reacted chemically with the zirconium in the fuel cladding and the pressure channel to form hydrogen.

Two explosions are known to have occurred. A possible explanation for the first explosion is that the rapid increase in the steam pressure initiated a steam explosion that ruptured the pressure channels. The exiting steam reacted chemically with the graphite to further increase the concentration of hydrogen. The hydrogen then reacted chemically with atmospheric oxygen, causing an explosion that blew the top off the reactor building. Another possible explanation for the second explosion was that it resulted from another steam explosion. After the explosions, the hot graphite was exposed to air and began to burn intensely, spewing radioactive materials from the fuel directly into the air. It took ten days to bring the fire under control.

More than 36 hours elapsed after the accident before the 135,000 people living within a 20-mile radius of the plant were told to evacuate. At the plant 2 workers were killed by the explosion and 29 by the overexposure to radiation, and 203 were hospitalized with radiation sickness. All the victims were plant operators, fire fighters, and other workers pressed into action to bring the fire under control. Of the 24 hospitalized victims who received bone marrow transplant operations to repair the body's immune system, 13 died.

The exact cause of the Chernobyl accident is not as significant as the fact that, as a result of the accident, large quantities of radioactive materials were allowed to disperse directly into the atmosphere. The first word that an accident had taken place came from Swedish scientists 750 miles away from Chernobyl who detected radioactive materials in fallout and concluded they could have only come from a nuclear reactor. Finally, two days after the accident, the Soviets acknowledged to the world that an accident had occurred.

As a result of the overexposure to radiation of the 135,000 people evacuated, it is projected that there will be about 500 additional cancer deaths in this population that would in any case experience about 27,000 deaths from all other kinds of cancer over the next 70 years. It is estimated that the total fallout of radioactive materials will also affect about 75 million people in the Soviet Union, of which 15 million are expected to die of cancer from all causes. The Chernobyl accident could increase the number of cancer deaths by 24,000 over the next 70 years. This number represents a very small statistical increase, about one additional death to each six hundred that are expected to die of cancer.

The accident caused direct economic losses of $2.7 billion to the Soviet Union. Also, the vegetation and water supply within a 20-mile radius of the Chernobyl plant remain highly contaminated with radioactive materials. A massive undertaking is required to decontaminate the area, make the food fit to eat, and the water safe to drink. The reactor must be entombed in a concrete case, heavily contaminated soils scraped up, and perhaps water processing plants built to purify the drinking water. Decontamination could take as long as four years.

As bad as the Chernobyl accident was, it still does not compare with the chemical accident that occurred in Bhopal, India, in 1984. That accident resulted in the deaths of more than 2,000 people, severe injury to more than 200,000 people, and long-term effects unknown at this time. That same year in Mexico City a liquefied gas facility exploded into a fire storm that killed more than 450 people, injured a far greater number, and left thousands homeless. In fact, almost every year, natural or man-made disasters occur that are comparable to the Chernobyl accident.

With the exception of countries adjacent to the Soviet Union that were subject to the initial radioactive fallout, the Chernobyl accident presents little risk to the health of people living outside the Soviet Union. In the 1950s, the United States and the Soviets tested nuclear weapons in the atmosphere that released large quantities of radioactive materials. The French and Chinese were testing nuclear weapons in the atmosphere until a few years ago. Obviously, radioactive pollution in the atmosphere is not desirable. However, the effect of fallout in the United States of these weapons tests and Chernobyl is negligible because the radiation increase is only a small fraction of the constant exposure we all receive.

Could a Chernobyl disaster happen in the United States? The answer is no. The Chernobyl nuclear plant would not satisfy U.S. design and safety standards and would not have received a license to operate in the United States. In fact, Chernobyl shows the soundness of U.S. decisions on reactor design, operation, and safety. Without a containment building, the Three Mile Island accident could also have had radioactive materials escape directly into the air instead of being trapped and retained in the containment building. Because of the lessons learned from Three Mile Island, U.S. nuclear power plants now have additional redundancy and diversity in all plant and safety systems as well as in emergency power systems. Much more rigorous training and requalification programs have been implemented for U.S. nuclear plant operational personnel to minimize human error and to assure proper response under emergency conditions.

9 Why Do Nuclear Power Plants Cost So Much?

It is difficult to comprehend how massive an undertaking is involved in constructing a nuclear power plant without visiting the construction site. However, a comparison to the World Trade Center in New York City should make for a better understanding of the complexities associated with building a power plant and the reasons why it is so expensive.

The World Trade Center is owned and operated by the New York and New Jersey Port Authorities. It consists of two 110-story office towers, two 9-story office buildings, an 8-story U.S. Customs House, and a 22-story hotel. The two office towers, each rising 1,350 feet, are the second tallest in the world. Some 50,000 people work in the World Trade Center. Another 80,000 come to the center daily.

The cost of the World Trade Center was about $1 billion in 1972. It contains more than 200,000 tons of steel. The 425,000 cubic yards of concrete used is enough to build a five-foot-wide sidewalk from New York City to Washington, D.C. Site excavation began in August 1966. The first tower was ready for occupancy January 1970 and the second tower in January 1972. At peak periods of construction, some 3,500 workers were on site at one time.

The construction of a two-unit nuclear plant also cost about $1 billion in the early 1970s. It contained about 270,000 cubic yards of concrete, 17,000 tons of reinforcement bars in the concrete, 11,000 tons of steel, 250,000 linear feet of large bore piping, 240,000 feet of small bore piping, and 6,700,000 feet of cable. The quantities of other bulk commodities such as instrument tubing, conduit, table trays, and valves are equally impressive. At peak periods of construction, the personnel requirement for engineering, construction, and operation was estimated to be about 4,000 workers. From ground-breaking to commercial operation, the project lasted about six years.

After Three Mile Island, the explosive growth in regulatory requirements increased the amount of bulk commodities by 50 percent or more. For example, the volume of concrete increased to about 450,000 cubic yards, stainless steel to 17,000 tons, and cable to 10,000,000 feet. The peak personnel requirement rose to more than 7,500 workers. The construction schedule increased to ten years and the total price exceeded $4 billion, with $1 billion associated with project financing costs.

To try to understand why the price of nuclear plants rose so fast, consider what would have happened if the New York City building code had changed radically when construction was nearly complete and most of the $1 billion for the World Trade Center had already been spent. Suppose it was mandated that much stronger protection was needed to protect against 300-mile-per-hour winds and the entire emergency plan had to be revised so that all 50,000 workers and 80,000 visitors in the buildings could exit twice as fast. Significant reinforcement of the foundations and other support structures to protect against wind damage would be needed. To satisfy the evacuation requirement, the emergency exits would not be sufficient, so the exit wells would have to be widened by several feet and perhaps an additional fire escape constructed outside the building. Since all these changes would have to be made on a nearly completed building, the cost in design time, materials, and craft personnel requirements would be very large, and the scheduled date for completion could extend a few additional years. In addition, if the Port Authority were to charge itself interest on capital invested in the total cost of the project, this cost would also be significant, comparable to the finance costs of a nuclear plant.

While ordering these modifications, if no consideration is given to a further increase of the total cost, the Port Authority could find itself in the unenviable position of making a decision to proceed with the changes and perhaps double the initial cost of the building rather than abandon the building and waste $1 billion. It also could find itself in a worse position of deciding to proceed with the changes based on assurances that it would not cost more than another $1 billion and then finding out the original estimates were incorrect or additional building code changes have made it impossible to complete the building for this amount of money. If in fact it were not possible to raise any more money after spending $2 billion, the project would have to be canceled with a total loss of $2 billion, an abandoned building, and an irate public.

No nuclear project has ever been canceled or a nuclear plant shut down because of unresolved technical or engineering problems. No plant has been canceled or shut down because it was deemed unsafe. The reason nuclear projects have been canceled is that the projected financial cost exceeded the ability of the utility companies to pay for them and/or the need for additional electricity in the near term was significantly less than was originally projected ten years earlier. If it had been possible to build these plants for $500 million per unit, the effect on the rate payer would have been minimal, and the excess capacity could have been used to retire older, less efficient plants and put coal-fueled plants in reserve.

10 Utility Management

The demand for a reliable source of electricity in very large quantities led to the formation of utility organizations that have the responsibility for constructing, operating, and maintaining power plants and the associated transmission and distribution systems. The United States has one of the best systems in the world for generating and maintaining a reliable supply of electricity to home and industry. This system also has sufficient reserves to satisfy large fluctuations in demand caused by dramatic changes in climatic conditions or unscheduled outages due to wind and storm damage.

There are a few hundred large utility companies and thousands of small utility companies in the United States. The majority of the larger utilities are investor owned and use the investors' funds to build power plants and transmission systems. They sell electricity to the rate payers, realize a profit on this sale, and pay taxes to local, state, and federal governments. The amount of profit is based on the return on investment, which is regulated by state utility commissions. There are also government-owned utilities, which are nonprofit organizations that range from very large utilities administered by federal and state agencies to very small municipal utilities administered at the local level. These government-owned utilities also sell electricity to rate payers to recover plant construction, operating, and maintenance costs. In addition, supplementary funds to cover costs are derived from other sources such as bonds and taxes.

Both investor-owned and government-owned utilities must project ahead to anticipate the industrial and residential growth in their service areas. A correlation exists between electricity demand and gross national product. Before 1970, the demand for electricity was growing at 7 percent per year, which meant that demand for electricity was doubling every ten years to satisfy industrial and residential requirements. There was no indication at that time that this rate of growth would not continue, so most utilities planned an aggressive power plant construction program. The choice of fuel was oil, coal, or nuclear. Oil was the fuel of choice. In fact, many coal plants were converted to oil rather than build additional auxiliary systems that greatly reduce sulfur emissions in their effluent to comply with environmental requirements.

Economic and environmental studies conducted by the utilities in the late 1960s showed that the total cost of generating electricity from nuclear plants was less expensive than that generated from either oil or coal and was far superior environmentally to coal plants. Also, coal had other inherent environmental problems such as mining, transportation, and waste disposal that were much more severe than could be anticipated for nuclear plants. In 1969, the Jersey Central Power and Light Company started operating the 620-megawatt Oyster Creek Nuclear Generating Station, the first plant that generated electricity at a cost that made it economically attractive for the utilities to look to nuclear power.

At that time, nuclear power seemed an excellent choice for future plants. Unfortunately, most utility managements made two major mistakes when they decided to build and operate nuclear power plants. First, they totally underestimated the technical and financial commitment that the utility had to make when it decided on a nuclear plant. Second, they based their project management decisions on their previous experience of building and operating fossil power plants.

Both of these mistakes can be attributed to a complacency that set in during the 1960s because of the successful financial and operational performance of the utilities. When it came to new fossil plant construction, most utilities would contract directly with an outside organization to design the plant, furnish major pieces of equipment such as steam generators, perform construction management, and carry out the construction. These utilities had a relatively small engineering group reviewing the construction schedule and budget. This system worked because the designers, contractors and subcontractors, and construction managers were proficient in their areas of expertise and were able to control costs and schedules with very little supervision from the utility company management.

When utilities decided to build nuclear power plants, they simply incorporated nuclear activities into their current corporate organization. Essentially all the upper management had little or no nuclear power plant experience. Most of the key positions related to nuclear plant construction and operation were filled with individuals who were qualified to build and operate smaller fossil plants. As a result, the nuclear project turned into a technical and financial nightmare. The typical utility management was unprepared for all the technical and environmental submittals necessary to obtain a construction permit. During construction, they were overwhelmed by the technical expertise of the organizations that design nuclear plants and manufacture nuclear equipment; they were confused by all the quality assurance and quality control requirements; and they did not appreciate the enormity of the effort to staff and train the operational personnel. In most cases, the utility was totally reliant on the expertise of outside organizations to obtain an operating license. In addition, an unsympathetic NRC was making major changes in design requirements and regulations that severely affected work already completed. It is no wonder that nuclear projects faced long delays and cost overruns.

In light of the high cost of nuclear plants, it may be hard to believe that essentially all utility managements are reluctant to spend money unnecessarily. However, as a group, they are extremely frugal and try to achieve a balance between keeping the cost of electricity low to the rate payer while providing a good return to their stockholders. It was extremely difficult for them to understand (because they had not done it before) that it was vital to their financial health to spend the money to hire large numbers of technical personnel experienced in nuclear power licensing, design, and construction at relatively high salaries to avoid cost overruns. Instead, most utilities, rather than spend money to strengthen their in-house technical capabilities, placed increased reliance on contractors who had little incentive to provide services in the most cost-effective manner. These contractors were being paid for their initial efforts as well as any changes or modifications to the plant. Recent NRC requirements have made utilities adopt organizational changes in which all utility nuclear activities are administered by a group with specialized expertise headed by nuclear-experienced personnel within the utility.

Added to utility management problems were the intervenors, the anti-nuclear activists who used most of the legal avenues open to them to delay the issuance of construction permits and operating licenses. Since finance charges amount to nearly one-third of the total cost of a nuclear plant, the delaying tactics significantly added to the total cost. It now is apparent that a strong public relations and public information program early on may have made many rate payers more sympathetic to the utility's problems and less interested in the rhetoric of nuclear power plant opponents. Such actions could have decreased the effectiveness of nuclear power plant critics and perhaps shortened the delays imposed by the courts.

Having been so badly bruised and battered in their nuclear power programs, utility companies will probably not place orders for new nuclear power plants until more certainty and stability have been introduced into the licensing and regulation process, construction schedules can be better defined, and there is greater public acceptance. Moreover, the excess capacity to meet peak loads may be adequate until the early 1990s. It is unlikely that there will be new orders for nuclear plants in the United States until after the year 2000. If and when new plants are ordered in the near future, they will probably be fueled with coal.

11 Nuclear Power Plant Insurance

Just like any other business, utilities need to have property and liability insurance to operate nuclear power plants. Property insurance premiums are based on the cost of the plant and can be acquired from commercial insurance companies. However, liability insurance premiums depend on operating and claims experience over a long period of time. In the 1950s little or no information on potential liability existed that could be assessed by private insurance companies because no accident had occurred at a nuclear power plant. As a result, the total amount of coverage provided by insurance companies was limited to $60 million for the entire industry. This amount was insufficient to allow utilities to proceed with the construction and operation of nuclear power plants.

In 1957, the U.S. Congress enacted a law, called the Price-Anderson Act, that required maximum available private insurance to be maintained, with the government obligating itself for an additional maximum coverage of $500 million. Thus, the total pool of funds to protect the public in the event of an accident was $560 million.

In 1975, the Price-Anderson Act was amended to phase out the government indemnity by assessing each licensed-to-operate nuclear unit $5 million in retrospective premiums. These premiums would only be paid if an accident occurred that exceeded the maximum available private insurance coverage, which had grown to $160 million. Thus, with 100 units now licensed for operation, the total pool of funds to protect the public is $660 million.

Before the Three Mile Island accident, no claims had been paid for injury to the public resulting from nuclear power plant operation. The Three Mile Island accident resulted in a large number of claims. Some of these claims were associated with loss of income and other expenses incurred due to evacuation of the area. Even though there were no appreciable radiation release and essentially no environmental changes, many claims were filed to compensate for mental anguish as well as illnesses, birth defects, and deaths that occurred at the time of the accident and were attributed by the claimants to the accident. All these claims were settled out of court by the private insurance companies at a total cost of about $40 million.

The Price-Anderson Act is in the process of being amended to increase the amount of private liability insurance coverage to about $200 million and raise the retrospective premiums that would be assessed to each licensed operating unit to about $60 billion. Thus, the total available pool for 100 units would be raised to about $6.2 billion with coverage being as much as $7.7 billion if 125 units eventually go into operation. Should an accident occur that exceeds the money allocated for this purpose, Congress has been empowered to vote additional funds.

12 Nuclear Power Plant Operator Training

The Three Mile Island experience proved that no matter how carefully a plant is designed to operate in theory, it is impossible to eliminate human error totally. The operators on duty at the time of the TMI accident were not stupid, but their analysis at a time of extreme stress was totally wrong. They were protecting against the wrong accident. Each of these operators had served in the U.S. Navy nuclear power program before joining the utility. They had been through what were believed to be rigorous training programs approved by the NRC and had been licensed by the NRC after passing with high grades the NRC's written and oral operator examinations.

The Presidential Commission investigating the TMI accident determined that the TMI operator training program conformed to the then existing NRC standards. It concluded, however, that these NRC standards allow a shallow level of operator training that did not adequately prepare operators to handle multiple failures. The Commission recommended more comprehensive training programs as a requirement for obtaining an operator license initially, and ongoing training programs to maintain the licensed operators' proficiency.

The lessons learned from the Three Mile Island accident have resulted in several improvements in the training and licensing process. The nuclear industry has funded a separate organization—the Institute of Nuclear Power Operations—that exercises quality control over all operator training programs and has established requirements for the accreditation of these programs. The NRC qualifications for operator applicants have been upgraded, and the passing grade on the operator licensing examination has been increased substantially. Also, the NRC has required that academic training programs contain more detailed presentations on heat transfer and thermodynamics and that they stress the fundamentals of reactor safety.

A major component in operator training is the extensive use ofnuclear power plant simulators. A simulator, which costs about $10 million, is a replica of the nuclear power plant control room, including all the plant instrumentation. Each plant system is described by mathematical equations programmed in real time on digital computers. All the plant parameters such as temperature, pressure, and power levels are calculated by the computer and recorded on the plant instrumentation. All plant operations from startup through full power to shutdown are faithfully reproduced. In addition, emergency conditions such as pump failure, inoperative valves, and pipe breaks can be simulated. For example, the scenario for the Three Mile Island accident can be recreated to make certain that operators take appropriate corrective action. Figure 8 shows a typical simulator.

The aircraft industry has been using simulators for pilot training and retraining for years. These simulator training programs ensure that the pilot maintains proficiency for such normal occurrences as takeoffs and landings and such abnormal occurrences as engine failure and landing gear malfunctions. The nuclear power plant simulator is similar in purpose to the aircraft simulator. By continual training and retraining on the nuclear simulator, operators will experience a large number of malfunctions and abnormal occurrences that prepare them to take appropriate action under emergency conditions.

We cannot eliminate human error totally in any complex technology designed and operated by man. The excellent safety record of the commercial airlines is attributable to the excellent training of the pilots. When aircraft accidents do occur, however, they are usually attributable to pilot error. Similarly, while nuclear power plant simulators greatly reduce the potential for human error, error will occur, but the probability of an error leading to a major accident should be very small.

The Apollo Space Progam that put a man on the moon had the highest level of test and evaluation for all equipment and systems aboard the spacecraft. Nevertheless, several Apollo space missions experienced serious equipment and systems failures. Appropriate emergency action by the well-trained crew was the difference between success and failure of these missions.

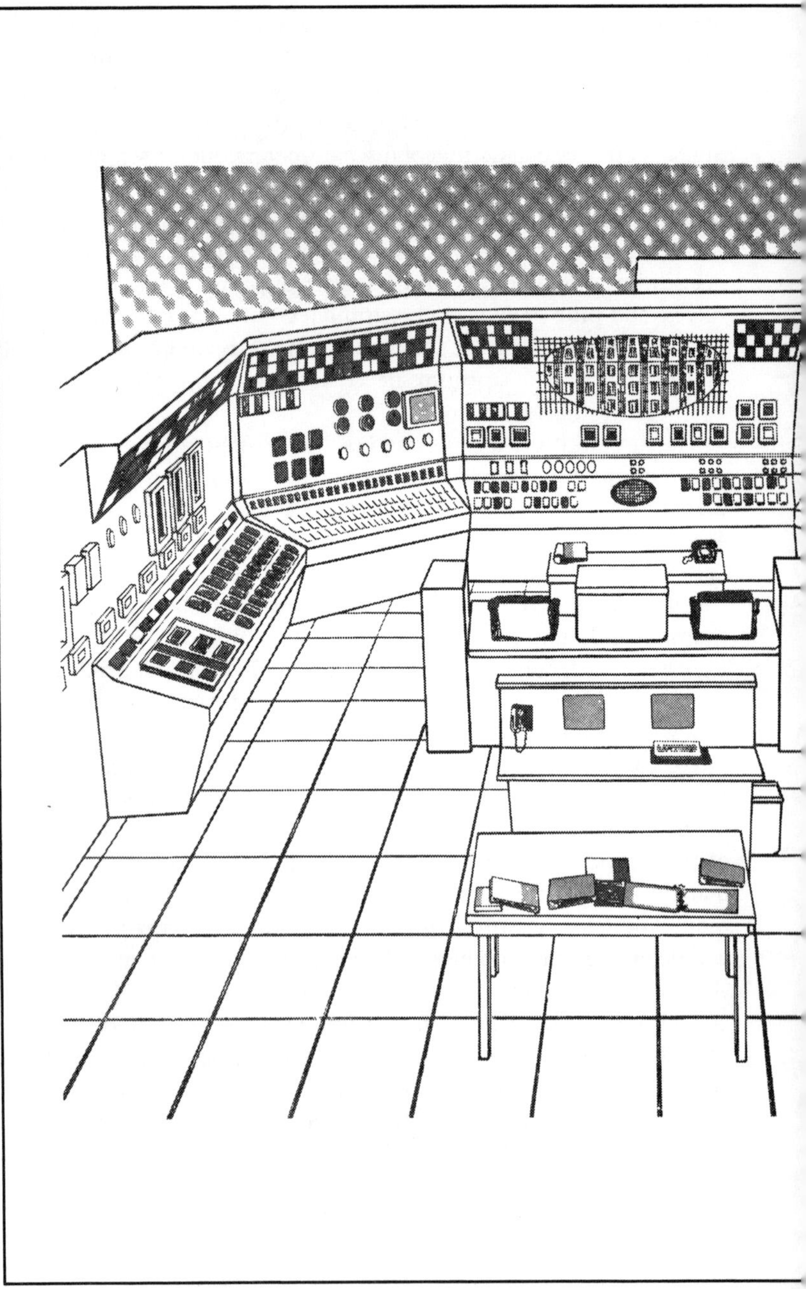

Figure 8. A nuclear power plant simulator

A nuclear power plant simulator is a sophisticated training device that is an ideal tool to train nuclear power plant operators. The physical arrangement of the simulator control room, including instrumentation, is identical to the control room in a nuclear plant. Each plant system is described by mathematical equations. A digital computer is used to solve these equations as a function of different plant conditions. The solutions to these equations activate the instrumentation in the simulator control room so that the readings are identical to the instrument response in the power plant control room. All plant operations from startup through full power to shutdown are faithfully reproduced.

The simulator control room has an instructor's console for setting all training instructions into the simulator. The various plant conditions are initiated from the instructor's console so that the operator trainees can gain the necessary skills in manipulating controls in response to specific operational situations, both normal and emergency. The operator thus acquires and retains the proficiency necessary for safe, reliable, and efficient operation of the plant.

No matter how much we automate power plant equipment, well-trained operators will always be an essential part of the overall safety system in nuclear power plants. For example, at one plant, workers accidentally set fire to the cables that provided electrical power to the nuclear plant systems. Damage to these cables prevented the normal use of many of the plant safety systems and radiation monitors. The plant operators shut down the plant manually and maintained the plant in a safe condition throughout the fire, which lasted several hours.

13 Risk Analysis

The public's reaction to participating in activities that pose a risk to life and health tends to be more irrational than based on actual statistics. It is well known that the use of seat belts could save thousands of lives each year and greatly reduce the severity of automobile accidents, yet many Americans refuse to use seat belts unless compelled by law. The fact that smoking cigarettes is hazardous to health and is a major contributor to more than 250,000 American deaths annually does not deter millions of Americans from smoking cigarettes. Alcoholic beverages consumed in excess amounts contribute directly to the deaths annually of more than 100,000 Americans and indirectly to the deaths of thousands in accidents. Relatively few people are known to be affected by radiation. However, to many, the imagined fear of radiation far outweighs the real dangers of death and illness associated with car travel, cigarette smoking, and consumption of alcohol.

Part of the problem is that the public does not understand the significance of statistics. Every heavy smoker will not contract lung cancer. In fact, many cases are cited where heavy smokers live very long lifetimes. However, statistics state that the odds are high that a heavy smoker will become a victim of lung cancer or some type of pulmonary or cardiovascular disease.

The degree to which we are willing to assume risks depends on the individual. When we purchase insurance, we are actually assuming that an unfavorable event may occur and we are protecting against severe economic consequences. Many people buy insurance when they fly a commercial airplane. On the other hand, most people are willing to assume the risk of an accident is small enough that the cost of insurance is a waste of money.

We do not normally lead our lives based on potential risk or we would all be totally immobilized. We do not forego electricity as a source of light because of the remote possibility that we may be electrocuted by a malfunctioning light switch. On the other hand, we do know that if we were to inadvertently come in contact with a high-voltage transmission line, the chance of survival is minimal, so it is a rational decision to avoid standing near fallen power lines. It is fair to say that the use of electricity is dangerous, but reasonable precautions, building codes, and regulations regarding its commercial use make the benefits far outweigh the risk.

In some activities, the risk of injury or worse is known to be high. Football and skiing qualify as sports likely to cause serious injuries. Piloting a private plane and skydiving are not covered by normal life insurance policies. Individuals engaged in known risky activities consider the benefit worth the risk.

Most activities have unknown or little known risks. The food we grow and eat are examples. The fertilizers, pesticides, and food preservatives that are now considered safe could be found detrimental in the future. The proper amount and type of such diet components as proteins, fats, and carbohydrates is subject to much controversy, and some advice being given today may prove to be in error tomorrow.

As an example, we do not consider a well-done hamburger as a potential source of colon cancer or as a cause of birth defects. However, recent scientific studies have shown that hamburgers broiled or fried well done contain six chemicals that cause mutations, chromosome damage, and cancer in rodents. Mutagens are compounds that change genetic material in cells and sometimes can cause cancer in animals. Sperm- and egg-cell mutations and damage to chromosomes, which determine inherited traits, can cause birth defects. On the other hand, eating undercooked meat can result in meat-borne diseases such as trichinosis and salmonellosis.

Based on the above information, it is doubtful whether the hamburger sales at McDonald's, Wendy's, or Burger King will suffer. We are all used to eating hamburgers, and it is difficult to believe that something as American as hamburgers could pose any personal danger. However, the fact that welldone hamburgers may adversely affect our offspring is another story. We may be willing to assume the risks, but we do not want to penalize future generations for our indulgences. If these findings are substantiated in human tests, we would probably all be more concerned about the condition of our hamburgers and may consume a lesser number.

Major problems associated with risk analysis are the source of the scientific information used as the basis of conclusions and the method by which the conclusions are presented to the public. The proper, or rational method, is to have recognized scientific experts carry out planned research programs or evaluate experimental evidence in a controlled and objective manner. The findings and conclusions from these efforts should be presented for publication in scientific journals where the entire process, including conclusions, is subject to evaluation by other scientists of equal stature before publication. In some cases, the submitted manuscripts may be published with little or no modifications; in others, additional information may be required before the manuscript is accepted; and in still others, the manuscript may be rejected because the data or methods used do not justify the conclusions.

In recent years, this process of evaluating experimental evidence has been circumvented by the print and television media. In their search for sensational news that for the most part serves to frighten the reader or viewer, the media presents information in an improper or irrational manner. Any individual with some credentials sufficient by media standards to be considered an "expert" is permitted to voice his or her opinions. In particular, the nuclear industry has been a victim of what should be considered to be biased media actions. An organization that uses a scientific name and advertises its objective to be the protection of the public against the "dangers" of nuclear power plants is considered to be an unbiased "expert," and its viewpoint is given instant recognition. No peer review is required before publication or presentation, and once published or presented its viewpoint is virtually impossible to refute by acknowledged experts.

The public would like to trust experts in making decisions on the risk associated with activities in which they participate. Most of the public accepts the judgment of experts that the use of seat belts, elimination of cigarette smoking, and decreased consumption of alcoholic beverages are in their best interest whether or not they choose to follow these courses of action. However, with regard to nuclear power plants, the information presented to the public has generated so much fear that to many, the risks of nuclear power outweigh the benefits. In addition, the fact that the choice of action is not controlled by the individual but by the government adds further to the public's uncertainty concerning nuclear power. It is hoped that experts will emerge in the future who can be trusted by the public and are acknowledged as authorities by the scientific community so that the public can make judgments about the risks of nuclear power based on fact rather than emotion.

14 Radioactivity and Radioactive Wastes

An atom consists of a nucleus of a positive charge surrounded by electrons of negative charge, the number of electrons being equal to the positive charge of the nucleus. If an atom is radioactive, it has the capability of transforming itself into another atom by the emission of positive, negative, or neutral radiation from its nucleus. If the resultant atom is stable, or nonradioactive, it is no longer capable of transforming itself into another atom. If the resultant atom is unstable or possesses the capability to transform itself into yet another atom, it will do so, and this process will continue until the end product is a stable or nonradioactive atom. Most atoms in nature are now stable. When the earth was formed many billions of years ago, most atoms were radioactive.

The process by which an atom transforms itself to another atom is called radioactive decay. The time it takes for this decay process to occur is described by a half-life. The half-life is defined as the time for half the identical atoms to decay, or be transformed, into other identical atoms. The half-life is a measure of the degree of stability of the atom. The shorter the half-life, the less stable the atom. Depending on the particular atom, half-lives can vary from fractions of a second to billions of years. A stable atom can be defined as one that has an infinite half-life or remains in the same configuration forever.

The earth contains huge quantities of radioactive materials such as uranium that will continue to emit radiation for billions of years. Uranium is found in nature in two forms. On an atomic weight scale, one form of uranium weighs 238, designated as U-238, and the other form weighs 235, designated as U-235. U-238, the most prevalent form of uranium, has a half-life of 4.5 billion years, meaning that in 4.5 billion years the radioactivity of the remaining U-238 atoms will be one-half of the current value. The decay of U-238 results in an atom that is radioactive, which in turn decays to an atom that is radioactive, with this decay chain continuing until the element lead is formed. Two well-known radioactive materials are a part of the U-238 decay chain: radium, which has a half-life of 1,600 years, and radon, which has a half-life of 3.8 days.

The energy source for the electricity generated in nuclear power plants is the heat produced primarily by the nuclear fission of U-235 and plutonium, Pu-239. The amount of U-235 present in the fuel is established in the fuel fabrication process. The plutonium is not present initially but is formed during reactor operation. The fission process results in the formation of two radioactive atoms, or fission products, which are also called radioactive wastes. These radioactive wastes have a wide range of half-lives. For example, a form of iodine, I-131, has a half-life of eight days, and a form of cesium, Cs-137, has a half-life of thirty years. If we were to store these two radioactive wastes over a two-month period, less than 7 percent of the I-131 would still be radioactive, but the Cs-137 would hardly be affected. Many radioactive wastes with long half-lives such as Cs-137 can be used as radioisotopic fuel to power orbiting satellites that transmit data back to earth.

If released directly to the atmosphere as occurred at Chernobyl, radioactive waste can be very hazardous to the surrounding population. In particular, I-131 and Cs-137 are particularly harmful. Iodine in the food we eat tends to lodge in the thyroid gland. Sudden ingestion of large quantities of I-131 would expose the thyroid to significant radiation. To protect against this type of exposure, pills are prescribed that contain natural iodine to saturate the thyroid. Then, if I-131 were to be ingested, it would pass directly out of the body and not affect the thyroid. Cs-137 can be absorbed in vegetation and thereby contaminate meat and milk products as well as vegetables and other crops. These commodities would have to be closely monitored for radiation to prevent long-term exposure in areas experiencing significant amounts of Cs-137 fallout.

Radioactive wastes are usually classified as being either high level or low level, depending on the intensity of radiation being produced by these wastes. Reactor fuel upon fabrication is a low-level source of radiation requiring minimum precautions against radiation. Upon its withdrawal from the reactor after having produced the designed amount of heat, the reactor fuel is a very-high-level source of radiation, and all handling and transporting of this fuel must be done remotely. The ultimate disposal of the fuel containing the high-level waste is governed by federal laws and regulations and is described in Chapter 18.

Natural levels of radiation found in dirt and water can be considered to be very-low-level wastes that do not require disposal. On the other hand, some 20,000 civilian enterprises use radioactive materials that emit low levels of radiation for a wide range of medical, industrial, and home applications. Also, the residues of water purification systems and items such as special handling materials at nuclear plants account for the majority of the total amount of low-level waste being accumulated annually. At the end of the useful life of these low-level radiation products and materials, their disposal is also governed by federal laws and regulations.

15 Radiation Sources and Measurements

Our environment contains many sources of radiation. Some of these sources are natural in origin, either originating from the sun or radioactive materials that have resided in the earth since its formation. Nature has mandated that we be exposed to these sources whether we like it or not. Other radiation sources are man-made, as illustrated in Figure 9. X-ray machines are used for dental and medical purposes, radioisotopes are used for medical diagnostic and therapeutic purposes, television sets are used for entertainment, microwave ovens are used to cook foods, and nuclear power plants are used to generate electricity.

Each of these radiation sources has somewhat different characteristics. It is possible to measure the biological effects of the different types of radiation in terms of a unit of radiation exposure called a rem. This is a relatively large unit, so the biological effects of radiation are usually described in millirem, which is a thousand times smaller. Since the discovery about 100 years ago that some materials found in nature give rise to radiation, an enormous amount of research has been conducted on the biological effects of radiation exposure and the methods to protect against overexposure. The technology exists for limiting exposure to man-made radiation sources. This technology is incorporated into the design and operation of nuclear power plants.

Unfortunately, in the minds of the public, the word radiation has become synonymous with the word hazard. Since every living thing on earth has been continuously exposed to radiation throughout its existence, radiation exposure should be considered normal. In fact, on the average, the exposure of every human being on earth to natural radiation is about 1 percent of a millirem per hour, or approximately 100 millirem per year. The problem is not eliminating radiation, which obviously can never be accomplished, but establishing radiation limits for man-made radiation sources to protect the public against overexposure.

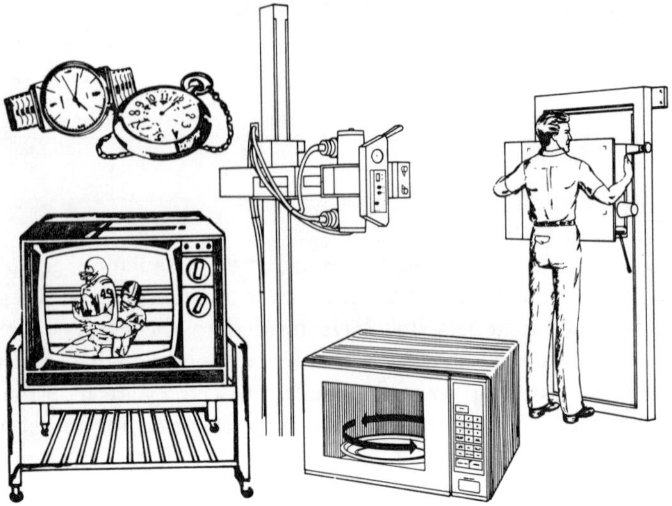

We have developed many radiation sources used in the home for illuminating dials, cooking, and entertainment, without posing a hazard to the public. Yearly medical and dental X-rays used for routine diagnostic purposes expose the average patient to about the same amount of radiation as is received from natural radiation. Without the use of X-ray technology to identify maladies at an early stage, a very large fraction of the population would be subject to serious discomfort; the average life span would undoubtedly be shortened. Low-level radiation associated with natural radiation and routine medical and dental X-rays results in a very small increase in cancer deaths and genetic birth defects. Damage to body cells by natural and synthetic chemicals in the diet and in the environment appears to be the major cause of most cancer and genetic birth defects and may contribute to aging and heart disease as well.

A nuclear power plant produces large quantities of radioactivity that remain within the plant, and only a negligible amount of radioactivity is released to the atmosphere. The radiation exposure of the public due to nuclear plant operation is insignificant when compared to the normal exposure resulting from natural and man-made radiation.

Figure 9. Man-made radiation

Natural radiation is a by-product of processes and materials created when the earth was formed and is necessary to preserve life. The sun on which we all depend for life-giving heat and light produces very highly charged particles called cosmic rays. We are bombarded continuously by these cosmic rays every day of our lives. Radioactivity from uranium and thorium and their by-products is found in rocks and soil. Because of the uranium and thorium in brick and granite, people living in brick homes and working in buildings having large quantities of granite are exposed to more radiation than people living and working in structures that do not use these materials. A serious problem that has only recently become known is the seepage of radioactive radon gas, a product resulting from the decay of uranium found in the earth. Radon penetrates through the foundations of well-insulated homes and subjects the residents to abnormally high levels of radiation. Inhalation of significant amounts of radon gas poses a serious health hazard.

Not many people realize that we all have radiation sources within our bodies, usually in the form of potassium, from which we receive about 25 millirem annually. This radioactive potassium constitutes a small part of potassium found naturally in the earth and is present in trace amounts in the food we eat and water we drink. This low-level radiation is unavoidable if we wish to continue the life process.

We all receive chest X rays, dental X rays, and other medical diagnostic radiation. These X-ray measurements identify potential medical and dental problems that cannot be observed visually. The average amount of radiation received by this means each year is nearly the same as that received from natural radiation, about 100 millirem.

The amount of radiation exposure from a nuclear power plant to the public is much less than that which the average person receives each year from natural radiation and from medical and dental X rays. If, for any reason, a plant is emitting radiation in excess of limits set by the government, alarms will sound and the plant must be shut down. As has been previously discussed, even under a major accident condition such as occurred at Three Mile Island, the surrounding populace is not subject to significant radiation.

Radiation measurement techniques are highly advanced and can detect concentrations much smaller than is possible with chemical separation techniques. Also, radiation instruments that monitor the environment can give immediate information on changes in radiation levels as well as identify the radiation sources. The realization that an accident had occurred at Chernobyl was based on measurements of increased radiation levels in Sweden 750 miles away. This ability to measure the smallest quantity of radiation protects the public from excess radiation. We are aware of its presence in the air, food, and water before it is ingested or inhaled by humans. As an example, a few years ago, as a result of an atmospheric detonation of nuclear weapons in China, there was a small amount of fallout near a U.S. nuclear plant. The slight increase in radiation exceeded the environmental limits by which the plant was licensed to operate and the plant was shut down because it was first believed that the radiation was coming from the plant. The state health department analyzed the situation and concluded the radiation levels were insignificant and would not contaminate the water or food supply in the area. This episode indicates that the radiation levels around nuclear power plants are well monitored and the maximum allowable levels are probably set too low.

Because it is easy to identify the presence of radioactive materials, necessary action can be taken in a short time. On the other hand, nonradioactive toxic materials can cause irreparable harm before their presence is recognized. The dumping of toxic wastes at Love Canal near Buffalo, New York, and the spraying of roads with PCB in Times Beach, Missouri, produced a wide range of health effects, including birth defects, on residents of these areas. It was not until years had passed, in some cases, that people found out they had been continually exposed to dangerous chemicals.

16 How Dangerous Is Radiation?

Given a population of 100,000, how many will die? The answer is everybody. We cannot predict with any certainty when and how each member of the population will die. However, based on statistical information relative to similar population groups, we can predict the average life span of this population and estimate what the final cause of death will be.

In the United States, approximately 2 million people die each year. More than 70 percent of these deaths result from infirmities that afflict the elderly, such as diseases of the heart, cancer, and stroke. We all have different life expectancies dependent on such factors as family history, occupation, area of the country in which we live, and life-style. Of every five deaths in the United States, one is from cancer. In 1984, about 452,000 Americans died of cancer, and approximately 65 percent of these deaths occurred after the age of 65. If we could totally eliminate cancer in the future by creating a pill or vaccine, the average life span would increase by only 2.7 years. Those people saved from cancer would die from some other cause. On the other hand, if through regular physical exercise and improved dietary habits, the incidence of heart attacks would decrease, the average life span would increase accordingly. Since there would be less death from heart disease, the number of ultimate deaths from cancer would probably increase. It is now estimated that 12 million Americans will die from cancer between 1970 and 2000, with 300,000 of these deaths being radiation related. In other words, less than 3 percent of cancer deaths are attributed to natural and man-made low-level radiation, and in excess of 97 percent of cancer deaths are not radiation related. These statistics indicate that while the consequences of low-level radiation exposure are significant, the expected incidence of cancer from natural and other causes is more than thirty times greater. Table 1 shows the predicted causes of cancer and resultant deaths between 1970 and 2000, including those caused by the operation of 125 nuclear power plants in the United States.

Table 1.
Predicted Causes of Cancer and Resultant Deaths

Causes of Cancer	Estimated U.S. Cancer Deaths 1970 to 2000
All causes except radiation	11,700,000
Radiation	
Natural	200,000
Medical X rays	100,000
Jet airplane travel	7,000
Weapons fallout	7,000
125 nuclear power plants in operation by the year 2000	40

Natural radiation exposure varies throughout the country. Because of the higher elevation and the presence of rocks containing radioactive materials, the increased exposure of Colorado residents to natural radiation amounts to the equivalent of about 4 chest X rays per year. From this information, one might infer that Colorado's cancer mortality rate would be higher than the rest of the country's. Yet, the opposite is true; the residents of Colorado had an estimated cancer mortality rate for 1979 that was 50 percent lower than the average. These data indicate that while there could be an increase in radiation-induced cancer, it is more than offset by the decrease resulting from reduced incidence of other forms of cancer. Also, compared to the rest of the country, each Colorado resident is expected to receive excess natural radiation equivalent to about 300 chest X rays over his or her lifetime starting with the fetal stage of pregnancy. However, there is no evidence of higher genetic damage in Colorado.

A large number of experiments have been carried out over the years with laboratory animals to determine radiation levels that can induce cancer. This information can be interpreted to estimate the potential effects on humans. In addition, statistics are being kept on Japanese who survived the atomic bomb attacks on Hiroshima and Nagasaki and were exposed to large levels of radiation. These statistics show that the immediate effects of the blast killed about 100,000 people. Estimates of the radiation exposure of the survivors could result in the "ultimate" death of an additional 500 to 1,000 people by radiation-induced cancer, "ultimate" meaning that there are other competing causes of death, including cancer from natural causes, that may prevail.

Unfortunately, similar experimental and statistical information is not available regarding most of the hazardous substances that affect our existence. If such information were available, decreasing the amount of low-level radiation exposure would probably have a very low order of priority relative to increasing our life span and improving the quality of life. As an example, the recent accumulation of experimental measurements and health statistics on cigarette smoking has demonstrated clearly to the public the real dangers of this activity.

In addition to health risks from inhaling and ingesting various substances and from accidents at home or at work, our life span is also affected by other factors, such as life-style, geographical location, and socioeconomic status. Professor Bernard Cohen of the University of Pittsburgh developed a technique for quantitative risk evaluation that covers a wide range of conditions. Various risks or conditions were compared based on a potential decrease in life expectancy. For example, if a person's remaining life expectancy at age 20 is 55 years and he is subject to an event that has 1 chance in 1,000 of being fatal, the average life decrease is expected to be 55/1000 years, or 20 days. Most risks are with us at all ages, and their effects must be added up over a lifetime to obtain the total decrease in life expectancy.

The typical life span in the United States is now 76 years for women and 71 years for men. Table 2 shows a number of different factors that decrease this average life expectancy, including the operation of 125 nuclear power plants in the United States.

Another area of much public concern is the belief that radiation can induce excessive mutations that lead to birth defects in current and future generations. Normally, about 4 to 6 percent of all babies not exposed to man-made radiation are born with varying degrees of congenital defects. It has been difficult to establish unequivocal information on the effects radiation has on the human embryo. It appears that a radiation level of less than 1000 millirem has little adverse effect on the fetus. Low-level radiation associated with natural radiation and diagnostic X rays is estimated to cause less than 1 percent of all genetic birth defects. However, to protect against any possible defects, diagnostic procedures involving radiation

Table 2.
Decrease in Life Expectancy from
Specific Risk Factors

Factors	Estimated Decrease in Average Life Expectancy
Male rather than female	5.0 years
Single male,age 20 (never marries)	6.0 years
Overweight by 30 percent	3.6 years
Cigarette smoking	
1 pack per day	7.0 years
2 packs per day	10.0 years
Heart disease	5.8 years
Cancer	2.7 years
City rather than rural living	5.0 years
125 nuclear power plants in operation	Less than 12 minutes

should be avoided during pregnancy. As for long-range effects from large radiation exposures, such as occurred at Hiroshima and Nagasaki, after 40 years the percentage of children born with birth defects in those two cities is no different than the rest of Japan and is comparable to the statistics compiled in other industrial societies.

It is a misconception that radiation from nuclear power plant operation is somehow different from natural radiation and can cause more severe genetic effects. Radiation is radiation. When a sex cell or an embryo is hit by radiation, the effect is the same whether the radiation came from a natural or a nuclear power plant source. Furthermore, the natural and synthetic chemicals used to preserve and color our foods and the industrial chemicals, pesticides, hair sprays, cosmetics, alcohol, cigarettes, and drugs we use are much more significant than radiation in causing genetic damage.

Controlled radiation exposures are used widely in medical treatment. The public may not be aware that millions of patients are the beneficiaries of diagnosis and treatment using X-ray machines and radioactive materials. Radiological procedures and nuclear medicine techniques are becoming more prevalent in our quest to extend our life span and increase the quality of life. Machines using radiation sources can perform full body scans or examine specific organs to provide the detailed information necessary to effect a cure. Nuclear medicine involves the intravenous injection of short-lived radioactive substances that diagnose life-threatening conditions such as blood clots in the lung.

We know enough about radiation to state unequivocally that unnecessary radiation exposure should be avoided. However, since low-level radiation is part of the earth's natural environment, we cannot avoid radiation exposure. Most scientists believe that the degree of risk associated with low-level radiation above natural radiation is very low. For certain medical activities such as radiology and nuclear medicine, the benefits of the diagnosis or treatment far outweigh the risk to the patient. For nuclear power plants, those who build, operate, and regulate the plants believe that the amount of radiation released to the atmosphere is insignificant and is well worth the benefit of an assured supply of electricity.

17 Nuclear Fuel and the Nuclear Fuel Cycle

The fuel for nuclear power plants is expensive. All the fuel materials must be specially prepared and fabricated to meet high quality assurance standards. The initial cost of a complete core of fuel elements for a nuclear reactor is about $60 million. In spite of this cost, because of the large amount of energy produced per pound of nuclear fuel, nuclear power plant fuel costs are significantly lower than those for a coal plant. If all the energy in one pound of uranium could be extracted by the fission process, the amount of energy generated would be equivalent to burning 1,300 tons of coal.

Uranium, the basic ingredient in nuclear fuel, consists of atoms of U-235 and U-238. Normal uranium contains less than 1 percent U-235, with the remainder consisting of U-238. U-235 will split when it absorbs low energy neutrons. As a result of this fission, energy is released as well as several neutrons that will produce fissions in additional U-235 atoms. Commercial nuclear power plants require that the U-235 content be approximately 3 percent of the uranium present. The procedure used to increase the concentration of U-235 relative to U-238 is known as enrichment and involves an extremely complex and costly industrial process. A possible alternative to the enrichment process, which is currently not used in the United States, is the addition of plutonium to natural uranium because plutonium has the same nuclear properties as U-235. The enrichment process is one step in producing nuclear fuel.

The entire process from mining ore through fabricating fuel elements to disposal of spent fuel elements (or high-level radioactive wastes if the spent fuel is reprocessed) is known as the nuclear fuel cycle, summarized in Figure 10. The first step is mining, and uranium currently being mined is found in concentrations of 2 to 20 pounds per ton of ore. The uranium is removed from the ore and concentrated by a milling process. However, before this uranium can be considered for use as reactor fuel, it must be refined to remove all impurities. The refining process results in the formation of a gas called uranium hexafluoride. This gas is sent to a government-operated gaseous diffusion plant for enrichment. After enrichment, the uranium hexafluoride gas is taken to a fuel fabrication plant where it is converted to uranium dioxide powder. This powder is then formed into cylindrical pellets about 0.3 inch in diameter and 0.5 inch high. A fuel rod is formed by stacking these cylindrical fuel pellets in thin-walled tubes of zircaloy, a zirconium alloy. A typical fuel rod contains a 12-foot-long column of stacked uranium dioxide pellets. The rods are clustered in groups called "bundles" or "assemblies" that are held together by top and bottom metal structures. Each fuel assembly can produce heat equivalent to 250,000 barrels of oil. A typical pressurized water reactor producing heat to generate 1,000 megawatts of electricity uses 200 fuel assemblies containing 100 tons of uranium dioxide.

During operation, about 67 percent of the U-235 is consumed, fission products are created, and plutonium is produced. Approximately one-quarter of the spent fuel assemblies are replaced each year with fresh fuel so that the average residence time of fuel in the reactor is about four years. After removal from the reactor, the spent fuel is highly radioactive and continues to produce heat. It is handled remotely by special fuel handling equipment under water. The spent fuel is removed from the reactor vessel and transferred underwater to an adjacent pool where it is retired for several years. During this time, the shorter lived radioactive fission products decay, and radioactivity is greatly reduced so that the fuel can then be shipped to a repository for permanent storage in specially designed shipping casks. These casks are made of steel and lined with lead. They range from 25 to 100 tons and may be designed to carry as few as one fuel element or as many as ten. The smallest casks can be transported by truck; overweight truck shipments (requiring special permits) can carry two- or three-element casks; and the largest casks must be shipped by rail. The procedures associated with the transportation of spent fuel should assure the public that such shipments are indeed safe. Without question, the shipment of toxic chemicals in tankers trucked by road or sent by rail presents a far greater hazard.

The present government plan is to bury the spent fuel in monitored storage facilities as discussed in the next chapter. If we do

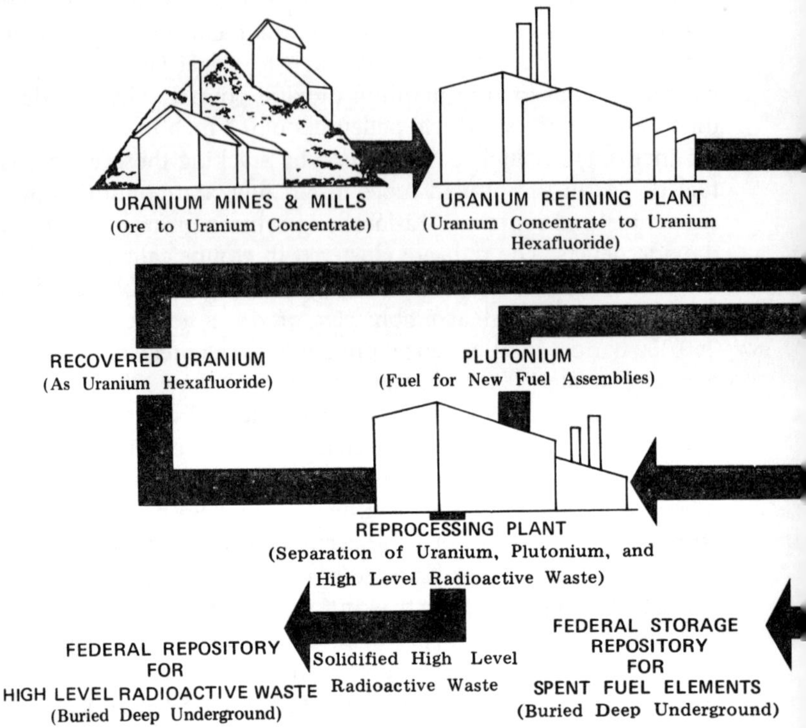

The nuclear fuel cycle involves several distinct processes starting with the mining of uranium ore and ending with the burying of the radioactive wastes that result from the fission of uranium and plutonium in the nuclear plant. After mining, uranium is concentrated and sent to a refining plant where impurities are removed and the uranium is converted to a gaseous form. This gas is transported to a gaseous diffusion plant where the U-235 content is enriched or increased to about four times the natural amount. The enriched U-235 fuel is shipped to a fuel fabrication plant where it is converted to uranium dioxide pellets to form fuel rods, which are grouped to form fuel assemblies. The fabricated fuel assemblies are transported to the nuclear plant and are placed in the reactor for about four years during which time they generate very large

Figure 10. The fuel cycle for commercial nuclear power plants

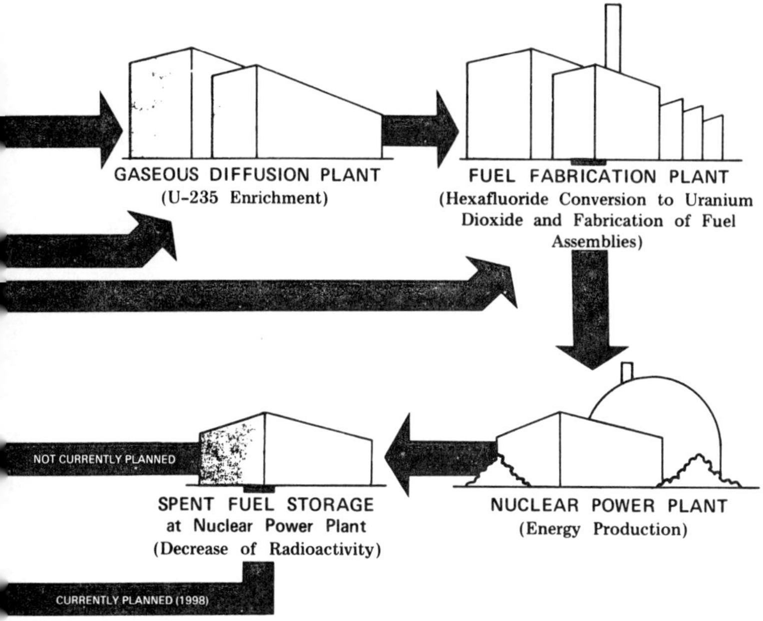

quantities of heat resulting in decreased concentration of U-235 and the formation of plutonium and high-level radioactive wastes. Upon withdrawal from the reactor, the spent fuel is highly radioactive and is stored at the nuclear plant for a period of years before shipment. Starting in 1998, it will be sent to a federal repository for storage. Since we would not be reclaiming the uranium and plutonium in the fuel for future use, this cycle is termed a throw-away cycle. If at some time in the future we decide to reprocess the fuel, the spent fuel is sent to a reprocessing plant where uranium is removed from the fuel and shipped back to the gaseous diffusion plant for reprocessing and utilization in fresh fuel assemblies. The plutonium can be separated and added in appropriate amounts to natural uranium in the manufacture of fresh fuel assemblies or, perhaps, used to fuel breeder reactors. The high-level radioactive waste would be solidified in preparation for burial at a federal repository.

not reprocess the fuel and reclaim the uranium and plutonium in the spent fuel for future use, this cycle is termed a throw-away cycle. Another option not currently available in this country is to send the spent fuel to a reprocessing plant where the high-level radioactive wastes are removed, solidified, and prepared for burial, and the fissionable material—U-235 and plutonium—is separated and reused to enrich natural uranium for use in new fuel assemblies. The plutonium can also be used to fuel breeder reactors that produce more fissionable material than they consume. The properties of plutonium are described in Chapter 20.

18 Disposal of Radioactive Waste

To dispose of high-level radioactive waste, the U.S. government intends to build repositories to store spent fuel from nuclear reactors. This program is embodied in the Nuclear Waste Policy Act of 1982, which mandates that by 1998 geological repositories will be in operation that are capable of storing 77,000 tons of spent fuel or 22,000 tons of solidified high-level waste and isolating the fuel or waste from the environment for 10,000 years. By law, only the federal government can develop and implement a program to dispose of radioactive waste. For more than forty years, the government has postponed taking any action on this important matter, giving rise to the mistaken impression that neither the knowledge nor the technology is available to dispose of radioactive waste in a safe and efficient manner.

Radioactive waste disposal is not a new problem created by the operation of nuclear power plants. For more than forty years, the government has been operating specially designed nuclear reactor facilities that use natural uranium fuel to convert uranium into plutonium for use in nuclear weapons. Reprocessing this fuel results in pure plutonium and the accumulation of high-level radioactive waste, primarily liquid, much of which is now being transformed into a solid form. The magnitude of the weapons program over all these years suggests that large amounts of high-level radioactive waste has been accumulated and that a permanent disposal site should have been designated many years ago. It is estimated that by the year 2000 the high-level waste accumulated by the U.S. weapons program will be almost ten times greater than the high-level waste being accumulated by the operation of all the commercial nuclear power plants. Thus, even without a nuclear power plant program in the United States, we would still need to have a high-level waste disposal program.

With all 125 nuclear plants now in operation or under construction generating electricity, it is expected that nearly 66,000 tons of spent fuel will be accumulated by the year 2000. This sounds like an enormous amount of waste until one realizes that U.S. industry produces 36 million tons of hazardous waste yearly and fossil power plants produce about 100 million tons of ash and sludge yearly. Over the next 15 years, industry and fossil power plants will produce more than 10,000 times the amount of hazardous waste by weight as will be derived from nuclear power plants.

All 66,000 tons of spent fuel containing high-level waste occupies a volume of about 10 football fields to a depth of about 14 feet. An area of 2 square miles, or about 2,500 acres, would be sufficient to house a repository where this spent fuel could be stored at depths of 2,000 feet to 4,000 feet below the ground and could be readily retrievable should the government decide in the future that reprocessing is necessary to reclaim the plutonium for the fueling of nuclear plants. To design and construct a monitored repository for spent fuel, the government has allocated $25 billion and will utilize more than 25 years of scientific investigation, laboratory analysis, and field experiments. Scientists and engineers are very cautious and conservative. One reason for the long delay in making decisions on a permanent repository is not disagreement over the fact that there is a way to dispose of high-level waste, but disagreement over the best way.

At the present time, three prospective sites for the repository have been chosen: Hanford, Washington; Deaf Smith County, Texas; and Yucca Mountain, Nevada. These sites have a stable geological formation, have proper hydrology conditions, and are far removed from economically useful resources. Conceptual designs for this facility include not only natural mechanisms for isolating the waste from the environment but a series of man-made barriers for containment. The overall facility is to be carved out of solid rock estimated to be 2,000 to 4,000 feet below the ground. The encapsulated spent fuel can be deposited in underground rooms with walls of thick steel. The spent fuel will be kept readily retrievable for up to fifty years in case we decide that we need it to fuel nuclear power plants in the future. As an added benefit, high-level waste from the defense program can use the repository and save the government about $1.5 billion. Once in operation, all areas around the facility would be closely monitored to assure that zero leakage is maintained. The 10,000-year criterion for a facility of this type is well within current technology. The pyramids in Egypt are several thousand years old, and caves have been found in France with hieroglyphics that indicate habitation more than 7,000 years ago.

It is somewhat ironic that we insist at this time that nuclear wastes be isolated for 10,000 years. We have no such restrictions on industrial dumps where highly toxic wastes are buried in containers that could corrode in a period of a few years. Because of their huge volumes, these toxic wastes now represent a serious immediate health hazard to our society. Also, in 10,000 years, essentially all the nonrenewable fuel sources to generate electricity, such as coal, should be totally depleted. There should be technological advances that will provide other fuel sources, but since the amounts of uranium will still be substantial and breeder reactors can supply an inexhaustible supply of fuel, nuclear power may be the primary source of generating electricity. Inhabitants of the earth at that time will probably consider radioactive waste disposal a normal problem and wonder why there was such a fuss around the year 2000.

For reasons difficult to understand, the popular media refers to any planned repository that stores spent fuel as a "dump" for "nuclear garbage." This wording projects an image of discarded junk randomly distributed and surrounded by a wire fence. A more correct picture is that of an intricate underground transit system built to exacting engineering standards. The facility is closely monitored to assure that radioactivity is isolated from the environment. It is also interesting to note that for more than two decades, the United States has carried out, and is still conducting, a number of weapons tests in Nevada that produce huge quantities of radioactive substances underground. This radioactivity remains underground and has little or no effect on the surrounding area above ground.

Another problem of lesser magnitude, but one that should be of great public concern, is low-level radioactive waste. This waste normally consists of a small amount of radioactivity distributed over a fairly large volume. About 60 percent of low-level waste is associated with the operation of nuclear power plants. The remaining 40 percent of low-level waste comes from 20,000 civilian licensees that use radioactive materials in medicine, research, and industry. Low-level wastes from all sources are ranked as Class A, B, or C depending on the half-lives of radioactive material. Nearly 97 percent of the waste is Class A, which consists of low-intensity, short half-life radioactive materials and requires a minimum of precautions for handling and packaging. About 2 percent of the waste is Class

B which must be stored for at least 300 years when it will have the same radiation level as natural soil. Less than 1 percent of the waste is Class C, which contains the highest quantity of long-lived radioactive materials and must be stored for 500 years.

The United States has only three active sites, located in South Carolina, Washington, and Nevada, to store low-level wastes. On the other hand, we possess more than 100,000 industrial waste dumps, 18,500 sites for the disposal of municipal solid waste, and 23,000 sewage-sludge sites. These sites contain nonradioactive chemical wastes, such as arsenic, barium, cadmium, chromium, lead, mercury, and selenium, which will retain their toxicity forever.

Federal legislation has been enacted to make each state responsible for disposing of its own low-level radioactive waste with rigorous milestones and stiff sanctions to assure that a sufficient number of additional disposal sites will be available in the future. Each state accepting the responsibility to regulate low-level disposal sites must adopt rules and standards compatible with federal requirements.

19 Nuclear Fuel Protection and Transportation

The nuclear weapons program has been in existence for more than 40 years. This country has more than 30,000 nuclear weapons. Extremely tight security measures prevent the theft of bombs and the possible diversion of strategic materials that can produce weapons. In the future, if we decide to reprocess nuclear power plant fuel, the same stringent precautions established for the defense program can easily be implemented for those parts of the commercial fuel cycle requiring such security.

A typical nuclear fuel assembly at a commercial nuclear plant is more than 14 feet in length and may weigh as much as a ton. It would be no easy matter for anyone to gain access to a nuclear plant and walk out with a fuel assembly. Outside the plant, spent fuel elements are shipped a few at a time in lead casks that weigh approximately 65 tons, making the theft or diversion of nuclear fuel during shipment before or after its use in the nuclear plant extremely difficult. Nevertheless, theft is not beyond the realm of possibility, so the NRC has implemented security precautions to ensure that losses of nuclear fuel will not occur.

Little would be gained by stealing a fuel assembly, either new or spent. It is impossible to construct an atomic bomb with nuclear fuel or any other nuclear material present at a nuclear power plant. For this purpose, it is necessary to use highly enriched uranium, or undiluted plutonium, which is never present at a nuclear plant. There is little motivation for a terrorist group to take over a nuclear plant and disperse the radioactive materials in the nuclear fuel for purposes of destroying a population center. Chemical and biological agents are more readily available that can be deposited into reservoirs for a far more lethal effect with much less hazard to the terrorists themselves. In fact, from a security point of view, nuclear materials are easy to detect because they constantly emit radiation. Electronically controlled security systems can instantaneously record the presence or movement of nuclear materials and actuate alarms when necessary.

20 Plutonium

Plutonium was present in the earth when it was first formed. However, since it is radioactive with a half-life very much less than the age of the earth, all of this plutonium has decayed to nonradioactive substances. Today, plutonium is formed in nuclear reactors. The most prevalent atomic form is Pu-239 which results from neutron capture in U-238. After its formation Pu-239 will fission when it absorbs low-energy neutrons in a manner similar to U-235. In some cases, instead of fissioning, Pu-239 converts to Pu-240.

Plutonium is a toxic substance. However, it is less toxic than many other radioactive materials, such as the radium and strontium used for medical and space power applications, and probably far less toxic than arsenic and cyanide, which are not radioactive. Plutonium handling methods and procedures are much more sophisticated than those for other dangerous chemicals, such as chlorine and arsenic. The United States has been producing, reprocessing, fabricating, and transporting tons of plutonium for more than 40 years without a single death attributed to plutonium poisoning. Also, radioisotopic power generators employing a nonfissionable form of plutonium, Pu-240, as a heat source are used to power communication satellites and space probes.

Worried about the possible diversion of plutonium for fabricating bombs, the government has halted all activities associated with the reprocessing of spent fuel and has ended the development of plutonium-fueled breeder plants. This policy has not deterred France, Britain, and West Germany from proceeding with the construction of reprocessing plants, nor has it lessened their enthusiasm for plutonium-fueled breeders. The threat has always been present that, instead of plutonium being stolen to manufacture crude bombs, high-yield bombs from existing world stockpiles could fall into the hands of irresponsible individuals or governments. Commercial nuclear power plants in this country do not add significantly to the magnitude of the problem. The total amount of plutonium that could be produced commercially by the year 2000 is only about 10 percent of the total amount produced in this country for the weapons program. Also, the commercial variety of plutonium is grossly inferior for bomb making because it contains a very large fraction of Pu-240, which reduces its explosive power.

Pu-239 has unique nuclear properties compared to U-235. Using plutonium in place of U-235 in a commercial nuclear power plant reactor that is cooled with liquid sodium instead of water can produce more plutonium than is consumed; this type of power plant is called a breeder. The resulting plutonium, upon reprocessing of the fuel, can be used to fuel additional breeder plants. The "tailings" from the uranium enrichment process used to increase the concentration of U-235 for commercial nuclear fuel have resulted in the accumulation of huge quantities of uranium that are depleted in U-235 content. By combining plutonium with the depleted uranium to fuel breeder reactors, we could produce electricity in the United States at the 1977 rate for 700 years. By 1990, the depleted uranium reserves will have increased sufficiently to permit an additional 1,200 years of 1977 electrical production. Without any additional uranium mining, the combination of plutonium and depleted uranium can provide this country with energy five times greater than the energy output obtainable from Middle East oil reserves and twice that obtainable from our total coal reserves.

Our current policy of not reprocessing spent fuel and prohibiting the use of plutonium in breeder reactors does not prevent other countries from developing the capabilities to fabricate nuclear weapons. Simpler methods are available to produce fissionable bomb material. India has already built and detonated an atomic bomb. Pakistan, South Africa, Israel, and Taiwan have the necessary facilities, which do not include commercial nuclear power plants, to join the nuclear weapons club.

The best example of differences in technology between the peaceful generation of electricity and the fabrication of bombs is illustrated by the People's Republic of China. The Chinese have been building and testing nuclear weapons for years but they do not have a single nuclear power plant in operation. The Chinese are now seeking technical assistance from other countries, including the United States, in building and operating nuclear power plants. It is expected that two small plants will be built and in operation by 1990.

To ensure world tranquility, effective international control and regulations that cover all sources of fissionable material should be established. A unilateral decision by this country not to use plutonium to produce electricity has little effect on the proliferation of nuclear weapons. In fact, if it were possible to convince the world's nuclear powers to reduce their stockpiles of plutonium for nuclear weapons by using plutonium to fuel commercial reactors, we could reduce world tension by literally turning our nuclear swords into electricity plowshares.

21 Cooling Towers

Nuclear reactors produce large quantities of heat, resulting in steam that drives a turbine generator. Only about one-third of the total heat is converted to electricity. As was shown in Figure 3, the steam leaving the turbine is condensed by a separate cooling system that circulates about one million tons of water per day to remove the remaining two-thirds of the heat. Nuclear power plant sites are located adjacent to rivers, lakes, and oceans to satisfy the requirement for very large quantities of cooling water.

The intake temperature of the condenser coolant water is colder in the winter than in the summer. Upon circulating through the coolant system, the return, or discharge, temperature of the coolant will increase by 10 to 15 degrees Fahrenheit compared to the intake temperature. An increase in water temperature is not necessarily detrimental to aquatic life. Fishermen usually find that the best fishing areas are in the vicinity of warm water effluent from fossil power plants. Warm water has also been found to accelerate the growth and increase the size of oysters and shrimp. However, an increase in water temperature during the summer months could decrease the water's oxygen content and precipitate a fish kill. In northern locations, the warm water may deter fish from migrating during the winter, and a plant shutdown at this time of year could also kill fish.

To avoid possible environmental damage, nuclear plants have auxiliary cooling systems that do not permit large volumes of cooling water to return directly into the source of coolant water. Instead, a variety of systems are used to recirculate the discharged water through a closed system and direct the heat taken from the power plant into the atmosphere. Most of these systems use cooling towers, as shown in Figure 11. Cooling towers are also used to control the specific thermal limits imposed on large fossil plants with regard to differences between the intake and return temperatures of the cooling water.

Cooling requirements are a major area of concern associated with the operation of large power plants. These plants have a low efficiency for converting heat into electricity, so as a part of normal operation, a great deal of heat must be removed from the plant. The result of this heat removal increases the temperature of the discharge water compared to the intake water. Federal regulations governing the water temperature of rivers, lakes, and bays require that power plant operation introduce relatively small changes in water temperature. To meet these regulations, many plants, both nuclear and fossil, are building cooling towers. Some of these cooling towers are very tall, 500 feet high and 400 feet in diameter at the base. Warm water from the plant is discharged into a perforated basin at the inside of the tower and falls into another basin near the bottom of the tower. During its fall the water comes in contact with the cooling air entering from the bottom of the tower and transfers heat to the air. The moisture in the air is condensed to steam, which flows out of the top of the cooling tower. Most of the water is returned to the plant for reuse. A small portion is discharged back to its source at nearly the same temperature it had before intake into the plant.

Figure 11. A cooling tower

Cooling towers are large and imposing structures with striking architectural lines. The television coverage of the Three Mile Island accident focused on the cooling towers because they dominate the site and almost ignored the containment building that housed the damaged reactor. As a result, cooling towers are now a symbol used by the news media to identify a story about nuclear power. Most cooling towers for nuclear plants are of a natural draft type. These towers are designed to have air enter at the bottom. The air interacts with the warm water discharged from the plant falling from about mid-height inside the tower. In this process, moisture in the air is condensed to steam that passes out of the top of the cooling tower. The discharged water is cooled to intake temperature, collected at the bottom of the tower, and most of it is pumped back to the plant coolant intake to repeat the cooling process.

22 Coal Power

The United States is fortunate to have a large abundance of coal. Our coal reserves are sufficient to provide fuel to satisfy all of our electricity requirements for several hundred years without using nuclear power plants. Unfortunately, the use of coal in very large quantities introduces problems that, over a long period of time, could create irreversible environmental damage not only to the United States but to the entire world.

The burning of coal in large power plants leads to the discharge of tons of gases, including sulfur dioxide and carbon dioxide. The sulfur dioxide forms sulfuric acid in the atmosphere. Sulfuric acid is believed to be the primary contributor to acid rain and the resultant destruction of aquatic life in lakes and streams, both near and far from coal plants. The carbon dioxide emitted to the atmosphere is creating a "greenhouse effect," meaning that energy from the sun that would ordinarily be reflected out of the atmosphere is retained. If these carbon dioxide concentrations are allowed to increase, the temperatures around the world would also increase, resulting in the melting of the polar ice caps and the flooding of seacoast cities. Current scientific estimates indicate that we could experience these adverse consequences in about 100 years.

The operation of a 1,000-megawatt coal plant is expected to cause 25 to 70 respiratory fatalities each year from the discharge of various types of pollutants, such as nitric oxide and mercury. Some of the materials discharged can produce highly toxic substances and be a source of birth defects. For example, the discharge of mercury leads to the formation of methyl-mercury, which is hazardous to humans and their offspring. The burning of coal also releases radioactivity to the atmosphere. Coal contains small concentrations of natural radioactive materials, such as uranium, thorium, radium, and radon. As a result of the combustion process, these materials are discharged from the stack. Licensing regulations would require that a nuclear power plant be shut down if it discharged a fraction of the radioactivity emitted by a comparable coal-burning plant.

Coal is a dirty technology. The mining, handling, and transporting of huge quantities of coal contribute to human suffering. Since we started mining coal underground in the United States, more than 100,000 miners have been killed in accidents. Because of the inherent dangers, Congress has undertaken the care of coal miners affected with black lung disease. The government is now spending more than $1 billion per year to compensate nearly 200,000 miners who have contracted this dreaded disease. Untold numbers are probably affected by coal dust in the vicinity of mines and by the shipping of coal in open railroad cars.

Major technological advances in coal plant design and operation and development of antipollution equipment have reduced the amount of pollutants discharged from the plant. Nevertheless, coal plant operations will always result in the emission of far more undesirable materials into the atmosphere than results from nuclear plant operation. Coal plants produce 100 million tons per year of ash and sludge in addition to residues from antipollution equipment. The total volume of these wastes is billions of times greater than the high-level wastes from nuclear plants. An unbiased assessment of the long-term problems associated with the disposal of wastes from coal plants would undoubtedly show that the storage of high-level wastes from nuclear plants has a simpler solution.

23 Alternative Sources of Electricity

Proven sources of large quantities of electricity such as coal and nuclear fuel have known technological, economic, and environmental problems. Proponents of unproven sources such as solar, wind, fuel cells, and nuclear fusion tend to emphasize the benefits of these sources and either neglect or do not know the many technical and safety problems that need to be solved before any of these unproven sources can be considered a viable option.

As an example, in the early 1970s, solar energy enthusiasts predicted that before the year 2000, 30 percent of the nation's energy would be derived from solar sources. Now it appears that at best solar energy will account for a small fraction of a percent of energy requirements by the end of the century. If we had gone along with the overoptimism concerning solar energy technology and decided to reduce our dependence on coal and nuclear plants, we would now be in real trouble. Even though sunlight is free, it has a very low energy density. The capital cost per installed kilowatt of a solar plant to convert solar energy to electricity by conventional means is at least ten times greater than a large nuclear power plant. These solar power plants may also have adverse environmental effects because of the very large land area required to collect and absorb a sufficient amount of sunlight. A more exotic system could use space technology to transmit a highly concentrated beam of sunlight to earth. The economics of such a design may also be exorbitant, and there could be a serious accident potential if the beam were somehow misdirected on earth or planes inadvertently flew through it.

Another approach to using sunlight to produce electricity is to employ photovoltaic cells that can convert a light directly into electricity. Many hand calculators use such cells as a power source. These cells are also used on space missions to provide necessary power for spacecraft. The cost of such systems measured in installed kilowatts is very high. A large decrease in production costs is necessary before such a conversion system based on photovoltaic cells can hope to compete economically with coal and nuclear power plants. Presumably, photovoltaic systems would power individual residences and factories directly. However, in urban areas with a high-density population living in large buildings, it is not possible to satisfy electricity needs with the limited amount of surface area available for direct conversion. Furthermore, when the sun does not shine, solar power becomes inoperative. Therefore, solar power will require a back-up system that uses coal and nuclear power plants to make certain that people do not freeze to death when it gets cold and are assured of light when it gets dark.

Essentially all American homes and factories now depend on electricity generated by large, centralized power plants. If it is possible to use an alternative technology to reduce our consumption of electricity, we should do so. An excellent example is the use of solar energy as a supplementary energy source to heat water. Furthermore, if a new technology can be developed that has demonstrated economic and environmental benefits, it can be phased in and our dependence on coal and nuclear power plants decreased. It should be understood, however, that to implement this new technology would require electricity generated by operating power plants. For example, if photovoltaic cells were to be developed and proved to be capable of satisfying a large portion of our requirements at a price we could afford, an enormous amount of electricity from conventional power plants would be needed to produce tens of millions of these devices. In other words, we cannot simply turn off the existing source of electricity should we choose to switch to an alternate source.

One truly exotic power source is nuclear fusion. It has the potential to satisfy our electricity requirements on earth forever. For the past 30 years, billions of dollars have been spent on research and development projects to design and build a machine that can duplicate the process by which our sun produces energy. A major advantage to a conceptual nuclear fusion power plant compared to a conventional nuclear power plant is that it can be designed to eliminate the possibility of a "runaway reaction" by limiting the amount of fuel that can be consumed. The obvious disadvantage is that, after all this time, we have yet to prove the economic and technological feasibility of fusion.

Nuclear fusion involves a reaction in which the nuclei of two light atoms interact to form a new atom and release a large quantity

Reprinted by permission.

of energy. This reaction is the source of solar energy. The atoms involved in the fusion process are various forms of hydrogen. Natural hydrogen consists of 99.985 percent of H-1, which has a mass of 1 on the atomic mass scale and 0.015 percent of H-2, known as deuterium in pure form, which has a mass of 2. Finally H-3, which is man-made, known as tritium in pure form, has a mass of 3. It is radioactive, having a 12.3 year half-life. The fusing of deuterium with tritium and deuterium with deuterium requires a minimum of at least 45 million degrees Celsius. All the nuclear fusion research over the past thirty years has been devoted to building a machine to sustain a reaction at these temperatures in which more energy is produced than is required to confine and ignite these atoms.

Assuming that it will soon be demonstrated that fusion can provide a large energy source, a sophisticated system to convert this energy to electricity must still be designed and built. Many safety and maintenance problems need to be solved before an acceptable plant can be put into operation. The handling of radioactive tritium is a major health and safety problem because it chemically acts like hydrogen, a highly combustible gas, and has to be transported to the plant. The amount of radiation produced by the fusion process will contaminate the entire plant, which could be the size of the Houston Astrodome, and all maintenance will have to be performed remotely. Thus, a nuclear fusion power plant, just like any other alternative source of energy, has its own risks that could prove to be far worse than a nuclear fission power plant.

24 The News Media

It is natural for all of us to be concerned about the nuclear power program in the United States. The nuclear industry has not distinguished itself in its attempts to gain the public's confidence. On the other hand, the news media have treated the industry unfairly. Since news reporting, whether by video or print media, is a highly competitive business, the tendency has been for news organizations to outdo each other to attract a larger audience and increase revenues. As a result, much of what is classified as news is really entertainment. Nuclear power has proved to be a particularly good subject because it can be used to generate fear and thereby capture the public's attention.

Most members of the news media lack an understanding of nuclear plant safety and health issues. Perhaps their own fears and anxieties are the reasons for presenting information about nuclear power in the most unfavorable way possible. As a result, the public has become more familiar with the technical difficulties and potential hazards of nuclear power than with its accomplishments. To illustrate the media's perspective on nuclear power, consider their treatment of two events that occurred about the same time as the Three Mile Island accident in 1979.

A serious train wreck occurred in central Florida. A large number of tank cars carrying highly toxic chemicals were derailed, and in the resulting explosion and fire, toxic fumes were released into the air. This accident required immediate evacuation of a 300-square-mile area involving more than 5,000 people. It was several days before people could safely return to their homes. The media devoted two days of prime coverage to this story. It did not bring forth any experts that insisted the public's safety required the shutting down of all railroads. In fact, the media reaction was only a mild rebuke of the railroad involved. No statements were made on the increased risk of developing fatal cancer. No statistics were presented predicting that the life expectancy of the survivors would be adversely affected, nor were statistics presented as to when the next train wreck would occur and how many thousands of people would be killed.

In Wichita Falls, Texas, a tornado killed about 50 people, left 20,000 people homeless, and caused more than $200 million of mostly uninsured damage. The media's reaction was one of sympathy for the survivors. The media tolerates an act of nature no matter how severe. There was no outcry for tornado-proof housing. There was no legislation proposed to vacate such a dangerous area permanently, since similar occurrences had happened before and would happen again. The media did not publish statistics as to when the next tornado would strike with the probability of killing more people and doing greater damage.

As a result of the Three Mile Island accident, no one was killed and the public was not exposed to an excessive amount of radiation. During the course of the accident, rather than try to alleviate public fear, the media blitzkrieg did everything in its power to indicate that we were witnessing the greatest catastrophe of all time. One member of the Presidential Commission created to investigate the accident, Anne D. Trunk, who resided within a few miles of the plant, described the media actions as follows: "Too much emphasis was placed on the 'what if' rather than 'what is.' As a result, the public was pulled into a state of terror, of psychological stress. More so than any other normal source of news, the evening national news reports by the major networks proved to be the most depressing, the most terrifying. Confusion cannot explain away the mismanagement of a news event of this magnitude."

The legal process related to the licensing of nuclear power plants allows for those opposed to nuclear power to be heard in an appropriate forum. Rather than simply report the results of these proceedings, the news media will often help to orchestrate unlawful demonstrations. By glorifying the actions of a small number of protestors who refuse to abide by the legal process, the news media encourages controversy and disrespect for the law. The media never questions the motivation of the protestors or asks where they derive their financial support. Extensive media coverage of planned unlawful demonstrations only serves to subvert the democratic process.

The nuclear industry consists of about 200,000 workers. As pictured by the news media, the leaders of the industry are shortsighted, greedy human beings whose only goal is increased corporate profits. They also tend to depict members of the industry as being unreliable and untrustworthy, whereas the few members of the industry who have joined the nuclear opposition are considered virtuous and honest. The viewpoints of politicians, consumer advocates, and environmentalists who are unequivocally opposed to nuclear power receive far more attention than the opinions of scientists who believe nuclear power should be a vital part of our energy future.

25 Nuclear Power in the Future

As long as it is believed that our electricity requirement can be satisfied by sources other than nuclear, it will be impossible to obtain the necessary government and public support to have a viable nuclear power program in the United States. If additional power plants are needed in the near future, the choice would have to be coal plants, even if we acknowledge that escalating the burning of coal could lead to an environmental disaster. Until there is a clear-cut decision by the government and the people that nuclear power is an essential part of the nation's energy program, no utility company would be willing to risk its future on the hope that a new nuclear power plant can overcome all the technical, financial, and regulatory obstacles that now exist.

Unfortunately, it is not possible to simply defer the nuclear option until a later day. Except for projects overseas, the companies that make up the nuclear industry have no justification to continue in a business in which they have no prospects. It took more than 30 years to build this industry, which is now wasting away. If we decide at some future time to build new nuclear power plants, we will have to import most of the technological and manufacturing capabilities.

Even though the rest of the industrialized world is investing heavily in nuclear power, we are being convinced that we can do without it. Our quality of life and high standard of living are based on a high level of technology that requires an adequate and reliable supply of electricity. Very few of us would prefer reading by candlelight rather than use electric lights that derive their energy from nuclear power plants. When the public becomes convinced that nuclear power is necessary, it will have recognized that nuclear power is a precious resource that holds the promise of providing all humankind with a limitless energy source.

More Books from GP Courseware

Nuclear Power Plant Steam and Mechanical Fundamentals, Second Edition

[illegible]

[illegible]

[illegible]

[illegible]